158
Advances in Biochemical Engineering/Biotechnology

Aims and Scope

This book series reviews current trends in modern biotechnology and biochemical engineering. Its aim is to cover all aspects of these interdisciplinary disciplines, where knowledge, methods and expertise are required from chemistry, biochemistry, microbiology, molecular biology, chemical engineering and computer science.

Volumes are organized topically and provide a comprehensive discussion of developments in the field over the past 3–5 years. The series also discusses new discoveries and applications. Special volumes are dedicated to selected topics which focus on new biotechnological products and new processes for their synthesis and purification.

In general, volumes are edited by well-known guest editors. The series editor and publisher will, however, always be pleased to receive suggestions and supplementary information. Manuscripts are accepted in English.

In references, Advances in Biochemical Engineering/Biotechnology is abbreviated as *Adv. Biochem. Engin./Biotechnol.* and cited as a journal.

More information about this series at http://www.springer.com/series/10

Lars J.C. Jeuken
Editor

Biophotoelectrochemistry: From Bioelectrochemistry to Biophotovoltaics

With contributions by

P.A. Ash · V. Fourmond · F. Hollmann · L.J.C. Jeuken · A.K. Jones · J.A. Laureanti · C. Léger · Y. Ni · M.M. Nowaczyk · N. Plumeré · K.A. Vincent

Springer

Editor
Lars J.C. Jeuken
School of Biomedical Sciences
University of Leeds
Leeds, United Kingdom

ISSN 0724-6145 ISSN 1616-8542 (electronic)
Advances in Biochemical Engineering/Biotechnology
ISBN 978-3-319-50665-4 ISBN 978-3-319-50667-8 (eBook)
DOI 10.1007/978-3-319-50667-8

Library of Congress Control Number: 2017938646

Printed on acid-free paper

This Springer imprint is published by Springer Nature
The registered company is Springer International Publishing AG
The registered company address is: Gewerbestrasse 11, 6330 Cham, Switzerland

Preface

The current drive towards green chemistry and sustainable energy has created an urgent need for light-driven catalysis, and bioelectrocatalysis has an important role to play. Photobioelectrochemistry or biophotoelectrochemistry defines a subfield that lies at the interface between bioelectrochemistry and photoelectrochemistry which is currently explored for its ability to contribute to green chemistry and sustainable energy. In this book we aim to the describe different aspects of biophotoelectrochemistry.

During the writing of this book, three clear applications of biophotoelectrochemistry became apparent and we have aimed to define the current state of the art of these three fields in the chapters "Biophotoelectrochemistry of Photosynthetic Proteins", "Artificial Photosynthesis: Hybrid Systems" and "Photosynthetic Microbial Fuel Cells". In the chapter "Biophotoelectrochemistry of Photosynthetic Proteins" we highlight systems where photoactive biomacromolecules such as photosystems I and II are incorporated in bioelectrochemical systems for electricity production. In the chapter "Artificial Photosynthesis: Hybrid Systems" we highlight the photocatalytic regeneration of enzyme cofactors and coenzymes for biotechnological applications. At the interface of the chapters "Biophotoelectrochemistry of Photosynthetic Proteins" and "Artificial Photosynthesis: Hybrid Systems" lie systems where light-harvesting entities such as quantum dots or dye-sensitised semiconducting nanoparticles are coupled to biocatalysts for solar fuel or electricity production. Finally, in the chapter "Photosynthetic Microbial Fuel Cells" we explore the latest field in biophotoelectrochemistry, namely microbial photoelectrochemistry, which can be defined as microbial electrochemistry of phototrophs.

When the chapters "Biophotoelectrochemistry of Photosynthetic Proteins", "Artificial Photosynthesis: Hybrid Systems" and "Photosynthetic Microbial Fuel Cells" were under construction, it quickly became apparent that the field of biophotoelectrochemistry relies heavily on previous advances in bioelectrochemistry. However, relatively few tutorial texts are available on the latter. To make this book accessible to a wide variety of scientists interested in the area, we designed three additional chapters to introduce the wider field of bioelectrochemistry. The chapter "Protein Electrochemistry: Questions and Answers" looks at the analysis and electrochemical methodology that has made bioelectrochemistry such a powerful tool. The chapter "Structure and Modification of Electrode Materials for Protein Electrochemistry" looks at the electrode materials commonly used in bioelectrochemistry (and biophotoelectrochemistry) and compares the benefits and drawbacks of the various materials. Finally, the chapter "Vibrational Spectroscopic Techniques for Probing Bioelectrochemical Systems" introduces spectroelectrochemistry, a physical chemical technique that allows the spectroscopic characterisation of biomacromolecules on an electrode surface. The latter is currently not yet commonly used in biophotoelectrochemistry, but its application will undoubtedly become extremely valuable to this field in years to come.

The areas that this book does not comprehensively cover are (non-biological) photoelectrochemistry and related fields such as voltaics. Many comprehensive texts are already available on these topics and we did not want to repeat previous excellent accounts of that work.

We hope that this book will provide an introduction to bioelectrochemistry and then continue with its application into biophotoelectrochemistry. The book has been designed to be useful both to experts in the area and to those from related sciences. It has been designed to have a forward-looking perspective where we introduce new techniques that are likely to become important in the field in the near future. We hope the book is enjoyable both as a self-study text and as a valuable review.

School of Biomedical Sciences
University of Leeds
LS2 9JT, Leeds, UK

Lars J.C. Jeuken

Contents

Adv Biochem Eng Biotechnol (2017) 158: 1–42
DOI: 10.1007/10_2015_5016

Published online: 16 September 2016

Protein Electrochemistry: Questions and Answers

V. Fourmond and C. Léger

Abstract This chapter presents the fundamentals of electrochemistry in the context of protein electrochemistry. We discuss redox proteins and enzymes that are not photoactive. Of course, the principles described herein also apply to photobioelectrochemistry, as discussed in later chapters of this book. Depending on which experiment is considered, electron transfer between proteins and electrodes can be either direct or mediated, and achieved in a variety of configurations: with the protein and/or the mediator free to diffuse in solution, immobilized in a thick, hydrated film, or adsorbed as a sub-monolayer on the electrode. The experiments can be performed with the goal to study the protein or to *use* it. Here emphasis is on mechanistic studies, which are easier in the configuration where the protein is adsorbed and electron transfer is direct, but we also explain the interpretation of signals obtained when diffusion processes affect the response.

This chapter is organized as a series of responses to questions. Questions 1–5 are related to the basics of electrochemistry: what does "potential" or "current" mean, what does an electrochemical set-up look like? Questions 6–9 are related to the distinction between adsorbed and diffusive redox species. The answers to questions 10–13 explain the interpretation of slow and fast scan voltammetry with redox proteins. Questions 14–19 deal with catalytic electrochemistry, when the protein studied is actually an enzyme. Questions 20, 21 and 22 are general.

Keywords Catalysis, Electron transfer, Protein electrochemistry, Protein film voltammetry, Voltammetry

V. Fourmond and C. Léger (✉)
Laboratoire de Bioénergétique et Ingénierie des Protéines, CNRS/Aix-Marseille Université, Marseille, France
e-mail: leger@imm.cnrs.fr

Contents

1 What Is the Difference Between "Electrochemical Potential," "Electrode Potential," and "Standard Potential"?

The term "potential" is used in electrochemistry to refer to distinct physical and chemical properties. It is important to understand the meaning of these different "potentials" to make sense of the Nernst equation (Fig. 1).

Before we consider redox equilibrium, let us recall the principles of chemical equilibrium. For a chemical system that contains two species, n_A moles of A and n_B moles of B, the total free energy of the system is $G = n_A\mu_A + n_B\mu_B$, where μ_X is the "chemical potential" of species X. If the system evolves according to the chemical equation $A \rightleftharpoons B$, then dn_A moles of A can be transformed into dn_B moles of B, and we define the reaction progress $d\xi = dn_B = -dn_A$, so that $d\xi$ is positive if the reaction proceeds forward, negative otherwise. The evolution of the system (at constant P and T) is such that its free energy, G, decreases, that is, $dG < 0$. This can also be written $dG = \Delta_r G d\xi$, where $\Delta_r G = \mu_B - \mu_A$ is the "reaction free

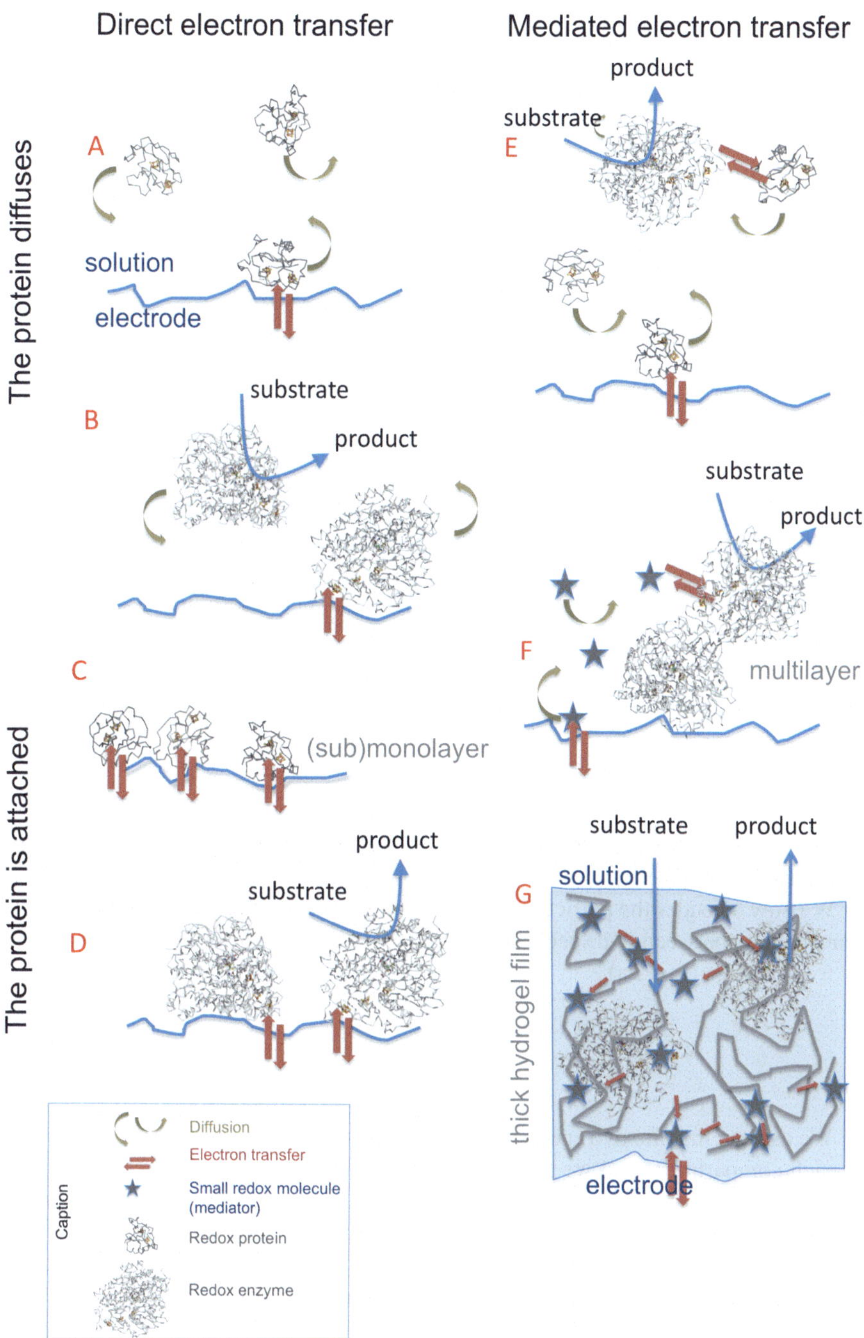

Fig. 1 A summary of various situations that can be investigated in electrochemistry, where the protein or enzyme is either free to diffuse in solution (panels **a**, **b**, **e**) or attached to the electrode (panels **c**, **d**, **f**), or entrapped and immobilized in a thick hydrogel film (panel **g**), and electron transfer between the electrode and the protein or enzyme is either direct (panels **a**–**d**) or mediated by either a redox protein (panel **e**) or a small redox molecule (panels **f**, **g**)

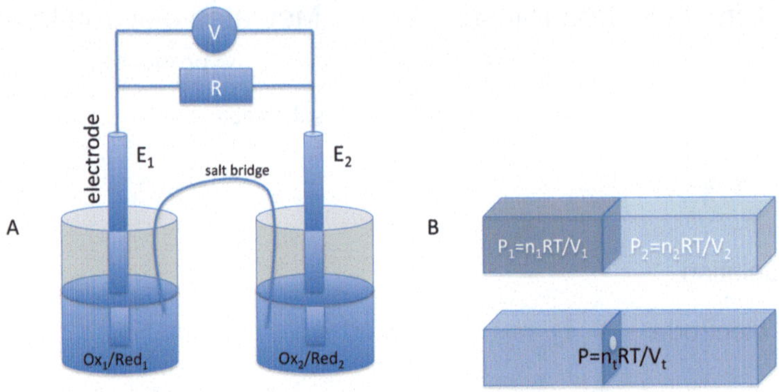

Fig. 2 A simple representation of a two compartment electrochemical cell (**a**) and an ideal gas analogy (**b**)

energy." The equilibrium is reached when $\Delta_r G = 0$, hence $\mu_A = \mu_B$. The chemical potential of a species depends on its standard chemical potential μ_X^0, and its activity, or, to make it simple here, concentration, C_X, according to $\mu_X = \mu_X^0 + RT\log C_X$, and the value of $\Delta_r G$ measures how far the system is from equilibrium:

$$\Delta_r G = \Delta_r G^0 + RT\log\frac{C_B}{C_A}, \tag{1a}$$

$$\Delta_r G^0 = -RT\log\frac{C_B^{\mathrm{eq}}}{C_A^{\mathrm{eq}}}, \tag{1b}$$

where $\Delta_r G^0 = \mu_B^0 - \mu_A^0$ and C_X^{eq} is the concentration of species X when the equilibrium is reached.

We now consider the particular case of a redox reaction that involves an electron transfer (ET) between two species that are in two different compartments connected by a salt bridge (Fig. 2a).

$$\mathrm{Ox}_1 + ne^- \rightleftharpoons \mathrm{Red}_1, \tag{2a}$$

$$\mathrm{Ox}_2 + ne^- \rightleftharpoons \mathrm{Red}_2. \tag{2b}$$

We assume that the two species can exchange electrons with an electrode placed in each compartment and we measure the difference between the electrical potentials of the two electrodes. If each compartment is under equilibrium, e.g., because a large resistance between the two electrodes prevents electron flow, the Nernst equation can be used to relate the equilibrium electrode potential, E^{eq}, the standard potential of the redox couple, E^0, and the concentrations of redox species in each compartment:

$$E_i^{\mathrm{eq}} = E_i^0 + \frac{RT}{nF} \log \frac{C_{\mathrm{Ox}i}}{C_{\mathrm{Red}i}}, \tag{3}$$

where n is the number of electrons involved in the reduction of "Ox" into "Red." If one is interested in the reaction that occurs in just one compartment, it is convenient to measure the electrode potential with respect to a "reference electrode" (AgCl, calomel, etc.) designed so that its potential is constant. Every electrode potential is measured with respect to a particular reference electrode, and the Union of Pure and Applied Chemistry requires that potentials are reported with respect to the Standard Hydrogen Electrode, using, e.g., $E_{\mathrm{calomel}} = 0.242$ V vs SHE [1].

The Nernst equation (3) is derived using the same reasoning that leads to (1a), except that, because the reactants have a charge and exchange electrons with a metal electrode, one needs to consider the "electrochemical potential" of each species, $\widetilde{\mu}_X$, defined as the "chemical potential" plus an electrical energy term $z_i F \Phi_i$ which depends on the charge of the species and the electrical potential of the phase that contains them (Sect. 2.2.4 in [1]).

$$\widetilde{\mu}_X = \mu_X + z_i F \Phi_i. \tag{4}$$

The electrochemical potential is therefore the partial free energy of a charged species; it is not something that can be measured using a digital voltmeter. The standard potential E^0 is a thermodynamic parameter that characterizes the redox couple. The greater the standard potential, the easier it is to reduce the species.

The system can be driven out of equilibrium either by using a potentiostat to set the electrode potential to a value that differs from the equilibrium electrode potential given by the Nernst equation or by decreasing the resistance between the two compartments and letting electrons flow. Electrons tend to go up the electrical potential gradient, and therefore electrons flow between the two compartments until the difference between the electrical potentials of the two electrodes is zero. The difference between the two electrode potentials is simply related to the free energy of reaction

$$\mathrm{Ox}_1 + \mathrm{Red}_2 \rightleftharpoons \mathrm{Red}_1 + \mathrm{Ox}_2, \tag{5}$$

by

$$\Delta E = E_2 - E_1 = -nF\Delta_r G, \tag{6}$$

and the standard potential of the reaction is related to the standard free energy of the reaction by

$$\Delta E^0 = -nF\Delta_r G^0. \tag{7}$$

It should be noted that ΔE and ΔE^0 have completely different meanings. The former is a physical property that is measured when one connects the two electrodes with a voltmeter. The latter is a thermodynamic quantity that can be read from tables or deduced from experimental results.

It may be useful to compare the above situation with the case of a perfect gas that would be compressed into two gas-tight compartments (Fig. 2b). The equilibrium pressure in each of the two compartments is given by the ideal-gas law (which is the equivalent of the Nernst equation). If one drills a hole in the wall that separates the two compartments, the gas particles flow from one compartment to the other until the pressure difference is zero and equilibrium is reached.

2 What About "Redox Potential" and "Reduction Potential"?

The parameter E^0 in Eq. (3) is called the "standard potential," "standard reduction potential," "reduction potential," or even "redox potential." We discourage the use of the latter because E^0 is related to a free energy of reduction (cf. Eq. (7)), and if it were a free energy of oxidation it would have the opposite sign. It does not really matter what we call E^0 as long as we know what we are talking about. Sometimes the term "formal potential" is used for the same quantity but only after non-idealities are taken into account. The "effective formal potential" refers to non-standard conditions, pH 7 for example.

3 And the "Current"?

The current has units of Amperes, which is electric charge per unit time (C/s). It is a measure of how fast electrons flow.

The current has several contributions. The Faradaic current corresponds to electrons crossing the metal/solution interface, as a result of electron transfer to/from species in solution. If it comes from the n-electron redox transformation between species "Ox" and "Red," then it is a measure of the *rate* of oxidation or reduction:

$$i = nFA\frac{d[\mathrm{Ox}]}{dt} = -nFA\frac{d[\mathrm{Red}]}{dt}, \tag{8}$$

where F is the Faraday constant (the charge of a mole of electrons, which equates 96,500 C) and A is the electrode surface (i/A is the "current density"). According to Eq. (8), a reductive (or "cathodic") current is counted as negative and an oxidation results in a positive current. The Americans often use the opposite convention for the sign of the current.

What makes electrochemical techniques so powerful at elucidating the mechanism of chemical reactions is that they can simultaneously tune the thermodynamic driving force (by changing the electrode potential) and measure the kinetics of the response.

4 What Is the "Overpotential"?

The term has two different meanings.

Overpotential can be a measure of the departure from equilibrium: the difference $E - E^{eq}$ between the electrode potential and the equilibrium electrode potential, so that the rate of the reaction is zero if the overpotential is zero. Sometimes one can consider the difference between E and E^0, which, strictly speaking, is not the same.

In contrast, in the context of electrocatalysis, Bard defines the overpotential as the value of $E - E^{eq}$ that is large enough that the rate of the reaction becomes above a certain threshold [2]. This overpotential is a kinetic property of the catalytic system. With adsorbed enzymes we discuss below that the magnitude of the current is irrelevant because the amount of adsorbed enzyme varies and affects the magnitude of the current. In that case, the overpotential is defined as the difference between E and the "catalytic potential," defined below, which is independent of the amount of enzyme that is adsorbed [3].

5 What Does the Electrochemical Set-Up Look Like?

The experiment is carried out using a potentiostat in conjunction with the cell (Fig. 3). The cell consists of at least one compartment (often two) and three electrodes. The electrode that interacts with the protein is called the working electrode. The potentiostat controls the potential difference between the working electrode and a reference electrode, which is often contained in a side arm linked to the main compartment by a capillary tip which is positioned close to the working electrode. To avoid passing current through the reference electrode, a third electrode, called the auxiliary or counter electrode, is used. It is often just a platinum wire. It is important to include in the cell solution a large quantity of inert "supporting electrolyte," often NaCl > 0.1 M.

Compared to many other biophysical techniques, electrochemistry is very cheap, with some potentiostats costing just a few thousand Euros. This may be one reason why the technique is used so widely.

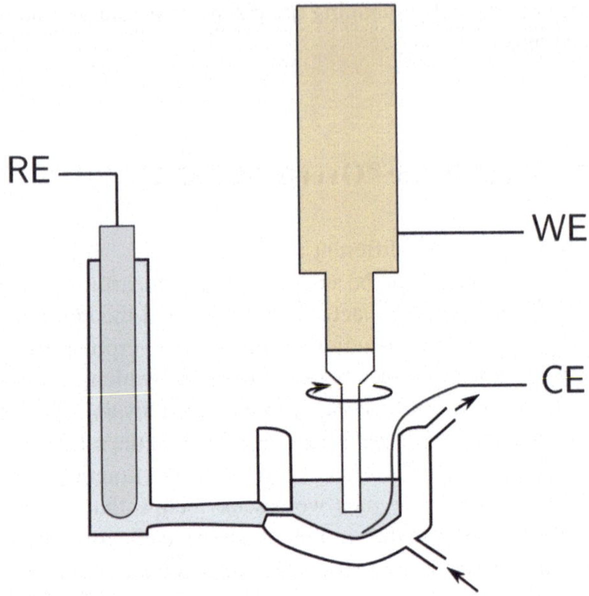

Fig. 3 A representation of a two-compartment electrochemical cell. *RE*, *WE*, and *CE* are the reference electrode, working electrode, and counter electrode, respectively

6 What Is PFV?

Protein film voltammetry (PFV) is the generic name for all electrochemical techniques that rely on direct electron transfer (ET) between a sub-monolayer of protein and an electrode (Fig. 1c, d) to learn about the properties of this protein [4]. In other techniques, a mediator is used to shuttle electrons between the protein and the electrode (Fig. 1e–g). The terms "protein film voltammetry," "protein film electrochemistry," and "direct electrochemistry of proteins" may be considered synonymous. Certain biofuel cells [5, 6] and third generation biosensors [7] are also based on enzymes directly connected to electrodes.

The "protein film" (Fig. 1c, d) consists of a very small amount of protein, less than a monolayer. The order of magnitude of the electroactive coverage (the number of enzyme molecules contributing to the electrochemical response) is therefore often lower than 1 pmol/cm^2 (here we refer to the geometric electrode surface), which corresponds to an average center-to-center distance between proteins of about 10 nm, on a flat electrode. The amount of protein that is needed to make the film may be much greater than that, because a large fraction of the sample is often "lost" in the process, but even so, it remains smaller than that required in the case of most biophysical techniques. In the most favorable cases this film is very stable and can be used for hours to acquire a large number of data sets.

The protein film is interrogated by measuring the current (which is proportional to the number of electrons being transferred per second) which results from the oxidation or reduction of the protein. As discussed in Sect. 1.10, in the absence of catalytic reaction (Fig. 1c), the change in redox state of all centers in the protein or enzyme as the potential is swept can be detected as a series of "non-catalytic" current peaks, the surface of which is proportional to electroactive coverage. When the protein is an enzyme, the electrode can substitute for the redox partner which would normally be used in a solution assay of the activity; in that case, in the presence of the enzyme's substrate (Fig. 1d), a steady current may be recorded, as ET continuously regenerates the redox state of the enzyme. The magnitude of this "catalytic current" is proportional to the turnover frequency and the electroactive coverage (unless the current is limited by mass transport; see below).

7 Can All Redox Proteins Be Studied By PFV?

No. The essential requirements are that (1) the protein can be adsorbed onto an electrode in such a way that direct ET occurs, (2) the amount of adsorbed protein is large enough that a significant current is measured, (3) the film is stable enough, and (4) the electrode does not alter the properties of the protein. In the case of an enzyme, an additional requirement is that the substrate and product of the enzyme are not electroactive on the electrode, otherwise their direct oxidation or reduction would interfere with the measurement.

Unfortunately, there is no way of predicting with certainty whether a particular protein can be studied. The situation is akin to that of crystallography. Even an experienced crystallographer cannot be certain that a particular protein crystallizes, that the crystals are large enough, and that there is no X-ray damage, before the results are obtained. An important difference is that many crystallography labs have robots that can be used to screen a large number of experimental conditions until a crystal is obtained, whereas attempts to adsorb proteins on electrodes are often not automated (see however [8]).

As a rule of thumb, the chances of success are greater with small and soluble proteins (but large membrane-bound complexes have also been studied [9]). It is essential that the protein has at least one surface exposed redox center, which can serve as an entry point for electrons into the enzyme; when this center is part of a redox chain that "wires" the active site to the redox partner, having it connected to the electrode effectively connects all the redox centers of the chain. In the case of enzymes, a signal can be measured even if the electroactive coverage is very small on condition that the enzyme has high turnover rate (the current is proportional to the product of the two), noting that the turnover frequency may be greater with the enzyme on the electrode than in solution (see, e.g., [10] for NiFe hydrogenase, and footnote 34 in [11] for nitrate reductase). If a certain protein has been successfully studied in PFV, it is likely that attempts to study a homologous protein succeed too (although the interaction with the electrode may depend on non-conserved surface-exposed amino acids).

8 What Is the Alternative to PFV?

Regarding electrochemical studies of enzymes, the majority of the literature concerns situations where electron transfer between the enzyme and the electrode is mediated, either by a small, non-biological, redox molecule, or by a small redox protein.

In a common configuration, the enzyme diffuses in solution and so does a smaller molecule (sometimes a redox protein [12]) which mediates electron transfer, i.e., shuttles the electrons to or from the electrode (Fig. 1e). The catalytic response of an immobilized redox enzyme connected to the electrode by a freely diffusing mediator (Fig. 1f) has also been discussed [13].

Alternatively, and as discussed in Sect. 1.5 and in [14] for photobioelectrochemistry, the enzyme can be immobilized into a thick film of redox hydrogel, a hydrated film of polymer whose side chains bear redox groups such as osmium complexes or viologen moieties (Fig. 1g). Electrons are transferred in the film by hopping, which is equivalent to a diffusion process (see Sect. 4.3.4 in [15]), and the substrate is also transported by diffusion in the film.

Mediated electron transfer is easier to achieve than direct electron transfer, but the latter is more desirable if the aim is to study the protein, because the amount of mechanistic information that can be gained is much larger. The fundamental reason for this is that the convolution between reaction and diffusion necessarily blurs the electrochemical response (and may add hysteresis), sometimes to the point that the electrochemical signal cannot be interpreted unless a specific and simple kinetic model is assumed.

9 How Can I Attach the Protein to an Electrode for Direct Electron Transfer?

Many methods have been proposed to attach proteins on electrodes, and this is thoroughly discussed in [16].

Various conducting (graphite, gold, silver) and semi-conducting (ITO) materials can be used. The choice of the material may depend on the reaction that the enzyme catalyses. It is important that, in the range of electrode potential where the enzyme is active, the electrode does not directly oxidize or reduce water or the substrate/product of the enzyme, else this would add a current contribution that has nothing to do with the enzyme and which may be difficult to correct. For example, graphite has been used with many redox enzyme that function under very reducing conditions, because graphite reduces water only at extremely low potential (much lower than the standard potential of the H^+/H_2 redox couple); hence graphite provides a flat baseline down to −1 V vs SHE or even lower. Under more oxidizing conditions (e.g., greater than 0 V for many copper proteins), gold may be a better choice, particularly because it is easily modified by forming self-assembled monolayers whose properties can be tailored to favor hydrophobic or electrostatic interactions with certain patches on the protein surface. Graphite is also easily covalently modified with carboxylic or amine functions that can form peptide bonds with amino acid side chains on the protein surface. Surface cystein residues (either natural or engineered) may sometimes be used for attachment on metal. Sophisticated methods are available for attaching membrane bilayers that embed enzymes onto electrodes, although in that case ET must be mediated by the quinone pool [17, 18].

Some attachment procedures are developed with the aim of selecting a particular orientation of the protein on the electrode, e.g., the orientation that gives a surface exposed redox center near the electrode surface. In the case of enzymes, there is often evidence for a distribution of rates of ET between the electrode and the enzyme, which could result from a distribution of orientations (see [3, 19] and the discussion of Fig. 9b, c, e, g, and h below). The redox properties of the redox center that is closed to the electrode may also be distributed, resulting in non-ideal peak shapes (broadening) [27, 28]. Kinetic and thermodynamic dispersions for proteins adsorbed onto electrodes have indeed been evidenced [29].

10 What Are Non-catalytic Voltammograms and What Defines Their Shape?

In most electrochemistry experiments carried out with proteins, one monitors the change in current resulting from a change in electrode potential E. The current response depends on how (and how quickly) E is varied, on the redox chemistry that is occurring, and on the transport of species from the bulk of the solution towards the electrode.

A very common procedure involves linearly sweeping the potential of the working electrode and measuring the current flowing between the working electrode and the counter electrode. When the electrode potential is swept forward and back, the technique is called "cyclic voltammetry." A voltammogram is a plot of the resulting current against electrode potential. A very important parameter in this experiment is the "scan rate" in units of V/s; the scan rate defines the time scale of the experiment, which should be compared to the response time of the process under investigation. With proteins and enzymes, experiments have been performed with scan rates ranging from a fraction of mV/s (in which case recording a single CV takes hours), to hundreds of V/s (to obtain information about reactions that occur on the ms time scale).

The current results from the ET on very short distances, to/from molecules that are a few Angstroms away from the electrode surface. Unless the sample is a redox protein that is adsorbed as a sub-monolayer film (Fig. 1c), "mass transport" (that is, the movement of molecules) to or from the electrode has to be considered. It is so if the redox protein freely diffuses in solution (e.g., Fig. 1a), or if we consider an enzyme adsorbed as a film, but which oxidizes/reduces substrate molecules coming from the solution (e.g., Fig. 1f). The redox reaction at the electrode surface can significantly decrease the concentration in the adjacent solution, resulting in the formation a micrometer-size "depletion layer." Mass transport results from diffusion, the Brownian motion which tends to homogenize the solution, and convection, the movement of molecules within fluids (migration, the response of a charged species to an electric field, is negligible in the presence of a supporting electrolyte). A common way to control and enhance mass transport is to force a convective movement towards the working electrode by rotating it. Typical rotation rates of 100–6,000 rpm (rpm = revolution per minute) are employed (see Sect. 15).

Whether you are looking at published data or running your own experiments, it is important to find out whether or not mass transport significantly affects the signal shape and magnitude.

11 How Can I Interpret Non-catalytic Voltammograms Recorded with Proteins at Low Scan Rates?

The non-catalytic current that results from the oxidation/reduction of a protein at a steady (i.e., "not rotating") electrode appears as one or a series of peaks of positive (respectively negative) current as the potential is swept up (respectively down). Table 1 shows the ideal characteristics of peaks resulting from the oxidation/reduction of diffusing or adsorbed species. Non-idealities always tend to make peaks broader, less intense, and more spread apart [27].

If we consider typical values of the relevant parameters in the case of proteins, a common feature of the two types of signals is that they are broad (≈100 mV) and small (because C, the concentration of protein in solution, D, its diffusion coefficient, or Γ, the amount of protein that can be adsorbed, are usually relatively small). Peak positions and peak currents are more easily measured than peak areas.

Table 1 Ideal characteristics (peak current, area, width at half height, separation) of peaks resulting from the reversible oxidation/reduction of diffusing or adsorbed species [1]. In the case of adsorbed species, "fast scan rate" means $v \gg RTk_0/F$, and k_0 has units of s^{-1}. For diffusing species, "fast scan rate" means $v \gg RTk_0^2/DF$ and k_0 has units of cm/s

	Diffusing species	Adsorbed species
Peak current (i_p)	$2.69 \times 10^5 n^{3/2} AC\sqrt{D\nu}$	$n^2 F^2 A\Gamma\nu/4RT$
Area	Not defined	$nFA\Gamma\nu$
WHH at low scan rate		≈ 90 mV/n
Separation at low scan rate	≈ 58 mV/n	0
Separation at fast scan rate ($n = 1$)	$RT/\alpha F \log(\nu)$	$2RT/\alpha F \log(\nu)$

C and Γ – bulk and surface concentrations of species
D their diffusion coefficient, n number of electron exchanged, ν scan rate, A electrode surface, α symmetry coefficient close to 0.5, k_0 rate of interfacial ET at $E = E^0$

Overlapping peaks are difficult to deconvolve, particularly if the background current is irregular.

The different situations described below are easily distinguished, e.g., in a log–log plot of i_p against ν (or, more often, by comparing plots of i_p vs ν and $\sqrt{\nu}$). The distinction is important because the interpretation of the peak current or peak area depends on whether the signal is proportional to ν or $\sqrt{\nu}$.

If the protein is adsorbed (e.g., Fig. 1c), the peak current is proportional to ν, because it corresponds to the oxidation or reduction of the entire sample. Since the charge that passes the electrode during a linear sweep is independent of scan rate, the current changes in proportion to scan rate (current is charge per time, which here is $\nu^0/\nu^{-1} \sim \nu$).

If the protein is free to diffuse in solution (e.g., Fig. 1a), the peak current is proportional to $\sqrt{\nu}$. The reason for this is that the current reveals the oxidation/reduction of species which are in a "diffusion layer" near the electrode, the size of which increases in proportion to the square root of time. Therefore, the charge that passes the electrode during a linear sweep is proportional to $\nu^{-1/2}$, and the peak current is proportional to $\nu^{-1/2}/\nu^{-1} \sim \sqrt{\nu}$.

An intermediate situation occurs with species that diffuse within a layer on an electrode. This is so, for example, if we consider the mediator moieties in the hydrogel film shown in Fig. 1g (they are attached to the polymer, but electron hopping between nearby redox centers gives the same behavior as a diffusion process). In that case, the dependence of peak current on ν changes from $i_p \sim \nu$ to $i_p \sim \sqrt{\nu}$ as the scan rate increases. This is because at fast scan rates only the redox centers that are near the electrode can diffuse towards the electrode and transfer electrons on the time scale of the voltammetry, whereas at slow scan rates the entire film can be reduced or reoxidized in a single sweep.

In all cases, the oxidative/reductive peaks should mirror each others, with (ideally) no peak separation if the species are adsorbed and a small (58 mV/n) peak separation if they diffuse in solution (Table 1). For adsorbed species, the area under the peak gives the charge passed for that redox couple (nF electrons per mole of adsorbed centers). Measuring the peak area therefore gives the surface coverage, which should be the same for the oxidative and for the reductive peaks, and the same for each redox center if a protein contains several.

As an example, Fig. 4b shows a non-catalytic voltammogram recorded with the protein azurin adsorbed onto a graphite electrode, which makes it possible to determine the redox potential of its copper site. In experiments, there is a capacitive contribution resulting from "electrode charging" (Sect. 1.12). The dashed line shows the interpolated capacitive current which has to be subtracted to obtain the contribution of the protein alone. Figure 4e shows the voltammetry of a cytochrome that diffuses in solution to/from the electrode. Figure 4c and f illustrate the response of adsorbed and diffusing proteins, respectively, which undergo two one-electron redox transitions. Note that in Fig. 4f, the two peaks have the same height. Note that in Fig. 4c, the two peaks have the same area, although the peak widths and heights are different. The CV in Fig. 6a can also be used to comment on the question of

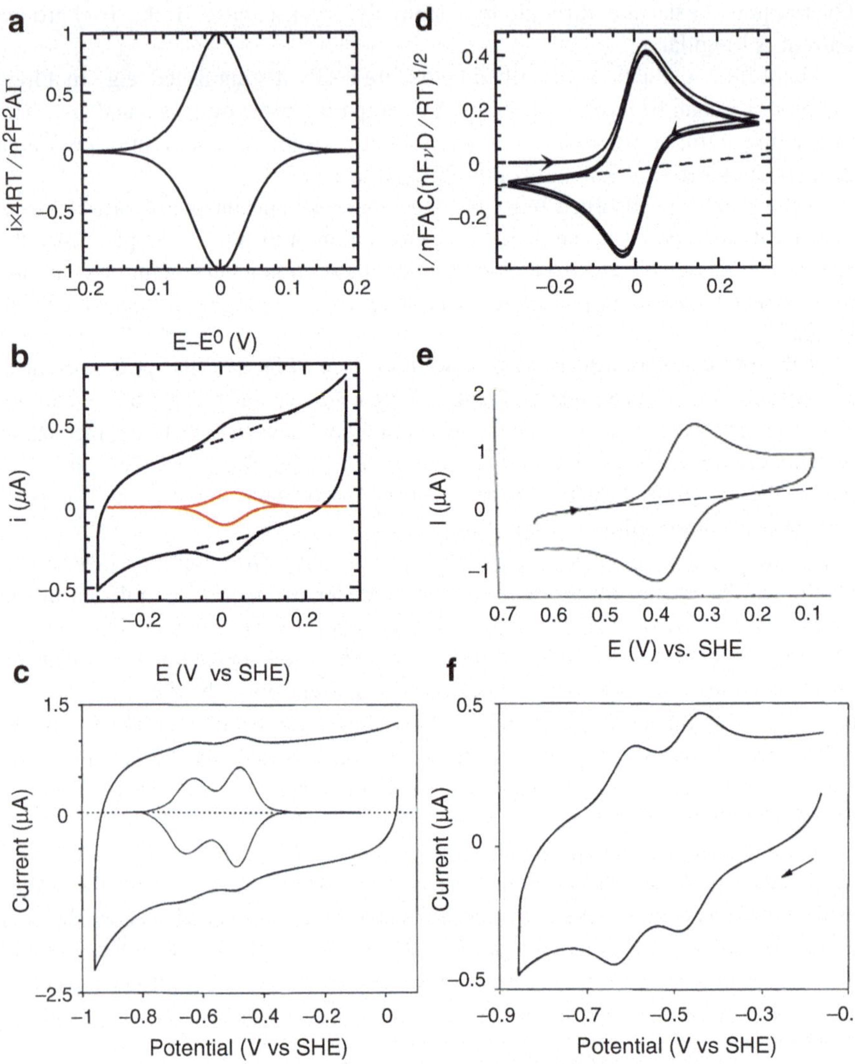

Fig. 4 Simulated (**a**, **d**) and experimental (**b**, **c**, **e**, **f**) non-catalytic, cyclic voltammograms for redox proteins undergoing direct electron transfer. The *left* column corresponds to the situation where the protein is adsorbed on the electrode, the *right* column shows data obtained with proteins that diffuse in solution. The *middle* and *bottom* panels show the responses for a protein containing one or two redox centers, respectively (hence the two *peaks* seen in the bottom panels). (**b**) Cyclic voltammogram for *Pseudomonas aeruginosa* azurin adsorbed at a pyrolytic graphite electrode. The *dashed line* is the baseline and the *inset* shows the baseline subtracted current (the Faradaic current). 0°C, pH 8.5. $\nu = 20$ mV/s. $\Gamma A \approx 5.5$ pmol. Data courtesy of Dr L. J. C. Jeuken. (**c**) Film voltammogram of a mutant of a ferredoxin of *Azodobacter vinelandii,* scan rate 20 mV/s. Reprinted with permission from [30]. Copyright (1998) American Chemical Society. (**e**) Data for diffusion limited voltammetry of the cytochrome c_2 from *Rps. palustris.* $C = 0.2$ mM, cell volume 0.5 mL (note the large amount of protein that is required for this measurement). Mind the axis! This voltammogram is plotted with the American convention: the high electrode potentials

stoichiometry, when a voltammogram shows several peaks. It has been obtained with a ferredoxin from *Azodobacter vinelandii* adsorbed onto a graphite electrode; this protein contains two FeS clusters (3Fe4S and 4Fe4S) that undergo three redox transitions. The high potential peak corresponds to the 1-electron reduction of the 3Fe4S cluster, from +1 to the 0 state. The low potential peak has two overlapping contributions: the one-electron reduction of the 4Fe4S cluster (+2/+1) and the two-electron hyper-reduction of the 3Fe4S cluster (0/−2). The area of the low potential peak is therefore three times that of the high potential peak. See [35, 36] for voltammetric studies of multi-heme cytochromes.

Generally speaking, if the scan rate is slow enough that interfacial ET can maintain Nernstian equilibrium for the redox species at the electrode surface, then the signal essentially informs either on the bulk or surface concentration of species and the standard potential of the redox transformation (obtained from the average position of the oxidative/reductive peaks). This may be valuable information, particularly when the redox center is difficult to titrate using conventional techniques. For example, see [37] for the electrochemical characterization of archaeal thioredoxins, and [38] for the detection of an auxiliary, low-potential 4Fe4S cluster in an enzyme of the radical-SAM family. If the protein sample is adsorbed, the electrode is easily transferred into a different solution, e.g., to examine the effect of pH. The measurement being very quick, it can be performed even under hostile conditions (e.g., in very acidic or basic solutions) before the enzyme denatures [39].

A reduction potential increases as the pH decreases because the reduced form of the redox center is necessarily less acidic (protons bind more tightly to the more negatively charged center), and increasing the concentration of protons therefore makes reduction easier. The number m of protons that are taken up upon n-electron reduction of a center is simply deduced from the slope of the linear change in $E^{0'}$ against pH: $-m/n \times 60$ mV per pH unit. A more detailed analysis of the pH dependence can yield the pK_as of the oxidized and reduced forms of the redox center [39]. Importantly, the same reasoning applies to the effect of any ligand that binds differently to the oxidized and reduced forms of a site, and whose concentration therefore affects the standard potential [20, 40].

12 What Is the Background Current and How Can I Subtract It?

In experiments where the potential is changed, whether it is linearly or stepwise, there is a background current that results from the electrode behaving as a capacitor.

Fig. 4 (continued) are on the *left*, and the current is positive for a reduction. Reprinted with permission from [31]. Copyright (1997) American Chemical Society. (**f**) Voltammogram of a solution of a mutant of a ferredoxin of *Azodobacter vinelandii*, 100 μM, scan rate 5 mV/s Reprinted with permission from [30] Copyright (1998) American Chemical Society

In voltammetric experiments, the background current is positive when the potential increases and negative when it decreases, and its magnitude is proportional to scan rate. Regarding voltammograms recorded with diffusive or adsorbed species, the intensity of the peak increases as the scan rate is increased (Table 1), but the benefit in terms of signal detection is modest or negative because the background (baseline) current increases in proportion to scan rate. Because in that case the contributions from the background current and the oxidation/reduction of the protein are additive, quantitative analysis requires that the baseline be *subtracted.*

In chronoamperometry experiments, where the current is recorded either at a constant potential, or in response to a series of potential steps, electrode charging is seen as a transient current whose magnitude decays exponentially after each step.

It is sometimes possible to subtract a baseline that is either recorded in an independent experiment with no protein or extrapolated from the part of the signal where the protein does not contribute (dashed lines in Fig. 4b, d, and e). Defining the baseline is easier in the case of adsorbed species, because the contribution of the protein is zero on either side of the peak(s), so that the current falls back onto the baseline (Fig. 4b).

In Marseilles we have developed two open source pieces of software that are particularly adapted to the analysis of electrochemical signals (www.qsoas.org) [41]. Most commercial, electrochemical softwares also embed baseline subtraction procedures.

It must be remembered that baseline subtraction is an endless source of artifacts.

13 What Can I Learn from Non-catalytic Signals Recorded at Fast Scan Rates?

In the limit of slow scan rate, the information that can be gained from non-catalytic voltammetry is only about the thermodynamics of the reaction ($E^{0'}$, acidity or dissociation constants, etc.). In contrast, data recorded at faster scan rates can give information about kinetics, because the redox process includes elementary steps (interfacial ET, protonation, and/or ligand binding or release, conformational rearrangement) that occur at a certain rate and delay the current response. Increasing the scan rate may shift the peak towards more extreme potentials or make it disappear entirely if the time required to record the CV becomes lower than time scale of the rate-limiting step. By examining how the signal depends on scan rate (or what the scan rate is that makes the peak disappear) it is possible to measure the rate of this step.

Section 2.2 in [15] summarizes how the voltammetric responses of species that diffuse and react in solution can be interpreted to learn about the *rates* of their transformations. We discuss below the electrochemical responses of proteins *adsorbed* onto electrodes because fast scan voltammetry is easier and more common in that configuration.

One example is the use of fast scan voltammetry to measure "the rate" of interfacial ET. We use inverted comas here because there is not a single value of "the rate" of interfacial ET, because the latter depends on the difference between E and E^0 according to

$$k_{\mathrm{ET}} = k_0 \exp^{\pm\frac{\alpha F}{RT}\left(E-E^0\right)}, \tag{9}$$

(+ for an oxidation, – for a reduction; α is a parameter that is often close to 1/2). What is often meant by "the rate of ET" is actually k_0, which is the rate of ET at $E = E^0$. Because rates of ET increase strongly (exponentially) as the overpotential increases, the ET reactions cannot be outrun by increasing the scan rate. However, the separation between oxidative and reductive peaks increases as ν increases (Table 1) because E passes so quickly across the value E^0 that, during the time needed to transfer electrons, the potential has shifted away from E^0, resulting in a peak that appears at a certain overpotential. A plot of peak potential against scan rate (on a log scale) is called a "trumpet pot" (Fig. 5b) and can be simulated with available software to determine k^0 (http://www.fbs.leeds.ac.uk/jeukengroup/freeware/index.php). A simple dimensional analysis suggests that, for adsorbed species, peaks separate when ν is greater than RTk_0/F (for adsorbed species,

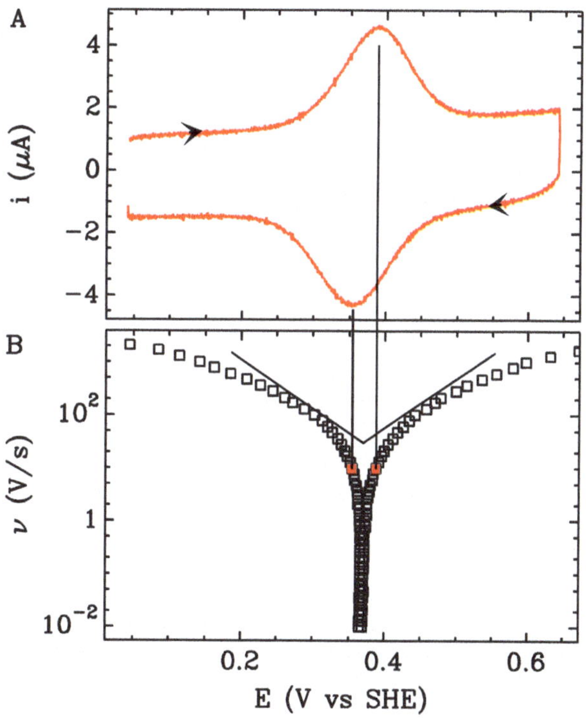

Fig. 5 Effect of scan rate on the voltammetry of a redox species undergoing a one-electron redox process. The data are for the "*blue*" copper protein, azurin, adsorbed on a modified gold electrode. (**a**) A cyclic voltammogram recorded at 10 V/s. (**b**) A "trumpet plot" of peak position against scan rate. Adapted from [42]. The *lines* show the ideal peal separation calculated (Table 1) for $\alpha = 0.5$ and $k_0 = 850\ \mathrm{s}^{-1}$ (the simulation in [42] of the complete data set returned $k_0 = 1{,}100\mathrm{s}^{-1}$)

k_0 has units of s^{-1}). For adsorbed proteins, values of k_0 as high as $1.4 \times 10^4\ s^{-1}$ have been measured [43].

Equation (9) is known as the Butler–Volmer equation. It is empirical, but Marcus theory applied to interfacial electron transfer predicts an exponential dependency of rate on electrode potential, provided the over-potential is small [1, 15, 44].

Fast scan voltammetry can also be used to learn about non-redox reactions whose rate depends on the redox state of the center. They are called "coupled reactions." An example is proton transfer, which occurs more favorably (lower pK_a, faster rate of protonation) in the reduced than in the oxidized state.

If protonation follows up reduction, and deprotonation precedes reoxidation, then there is a range of scan rate where, starting the CV from the high potential limit, the center can be reduced and protonated during the sweep towards low potential, but deprotonation is not fast enough to proceed during the sweep back to high potential, reoxidation does not occur, and the reoxidation peak vanishes. It is said that the deprotonation "gates" the reoxidation [33]. Figure 6 shows a series of CVs recorded at increasing scan rates, always starting from the high potential limit, with a ferredoxin mutant adsorbed onto an electrode [33]. Figure 6a shows the response on a large potential window. Figure 6b–f focuses on the high potential peak and how it changes when the scan rate increases. This peak is pH-dependent and corresponds to the proton-coupled reduction of a 3Fe4S center. The reduction peak is visible at all scan rates, but the reoxidation peak disappears at scan rates above 1 V/s, showing that deprotonation occurs on the time scale of seconds.

These methods can be used to determine the rate of any reaction that is coupled to ET (see [40] for an example about the rate of binding/release of imidazole to a copper center).

14 What Kind of Signals Can I Expect with an Enzyme?

If the protein investigated is a redox enzyme, then two kinds of signals can be detected.

Non-catalytic signals, obtained in the absence of substrate, result from the mere oxidation/reduction of all the redox centers in the enzyme. In the case of enzymes, these signals are rarely detected (see the non-exhaustive list of references [20, 45–52]). They are small when the enzyme diffuses towards the electrode because the diffusion coefficient of large proteins is small, and small when the enzyme is adsorbed, because the current is proportional to the electroactive coverage, which is low.

Catalytic signals result from a continuous electron flow between the electrode and the enzyme, the redox state of which is continuously regenerated by the reaction with the substrate. Let us consider oxidative catalysis, and compare the "diffusing," "adsorbed," and "hydrogel" configurations (Fig. 1b, d, and g, respectively).

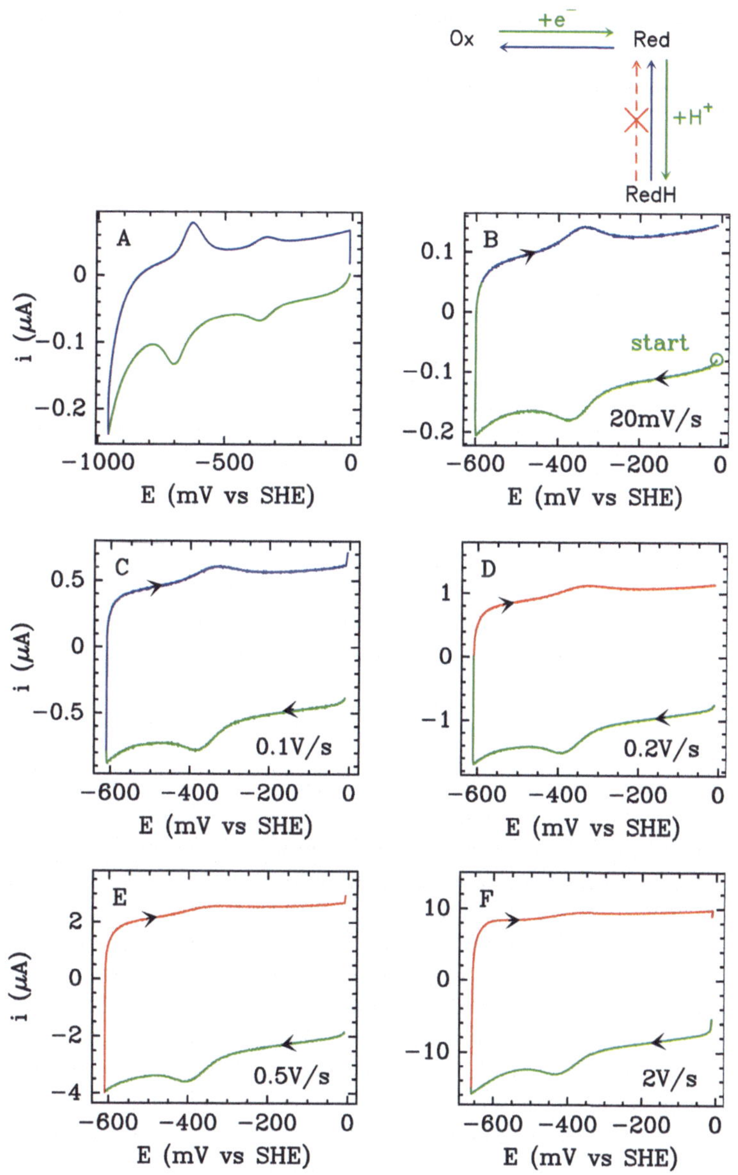

Fig. 6 Effect of scan rate on the voltammetry of a redox species undergoing a one-electron, one-proton redox process. The data are for the $[3Fe4S]^{+/0}$ redox couple of a slow proton-transfer mutant of Azodobacter vinelandii ferredoxin I (D15E) at low pH [32, 33]. Panel (**a**) shows a CV recorded on a large potential range. Panels (**b–f**) show CV recorded around the potential of the 3Fe4S cluster, at scan rates between 20 mV/s and 2 V/s, as indicated. Adapted from [34]. The videos available at http://mov.bip06.fr/playlists clearly illustrate how changing the scan rate changes the time scale of the experiment and the shape of the signal. The *colors* indicate which reaction occur (*green*, *blue*) or do not occur (*red*) depending on scan rate, when the potential is swept across the potential of the 3Fe4S center

If the catalyst, whose bulk concentration is c_E, is oxidized near the electrode (Fig. 1b), it diffuses away from it with a diffusion coefficient D_E, and is reduced in solution upon reaction with the substrate in a pseudo first-order process, then a steady state can be reached, where the concentration of oxidized enzyme decreases exponentially as a function of the distance to the electrode with a characteristic length that equates to $\sqrt{D_E/\mathrm{TOF}}$, and the catalytic current is

$$i_{\mathrm{cat}} = nFAC_E\sqrt{\mathrm{TOF} \times D_E}, \tag{10}$$

where TOF is the turnover frequency and n is the number of electrons produced in one turnover.

The interpretation of catalytic signals obtained with adsorbed enzymes (Fig. 1d) is somehow simpler. Let us consider oxidative catalysis. If the electrode potential is high enough, then the enzyme is oxidized, it oxidizes the substrate, and electrons are transferred to the electrode. This is measured as a current (charge passed per second) that is simply proportional to the TOF and to the electroactive coverage Γ_E.

$$i_{\mathrm{cat}} = nFA\Gamma_E \times \mathrm{TOF}. \tag{11}$$

If the enzyme has a reasonably high turnover rate, then catalytic signals for adsorbed enzymes can be detected even if the electroactive coverage is low. Compared to solution assays, direct electrochemistry with an adsorbed enzyme provides an additional control parameter, the electrode potential, and makes it possible to sample the activity at a high rate. It is easy to measure the catalytic current ten times per second. (Achieving greater time resolution is unnecessary because when the activity changes in response to a potential step, the initial transient is dominated by the capacitive response of the electrode, and when the activity changes in response to a change in buffer concentration, there is a dead-time corresponding to mixing [53].)

The "hydrogel" configuration (Fig. 1g) is the most complex because the transport and reactions of various species within the film have to be considered. Mass transport in the solution *outside* the film must also be taken into account unless the electrode is spun quickly. The current equation depends on which step is the slowest: catalysis, electron diffusion, or substrate diffusion in and outside the film. If we ignore mass transport outside the film, the parameters that characterize the system are the TOF (or more precisely the two Michaelis parameters k_{cat}, K_M and the rate constant for the bimolecular oxidation of the enzyme by the redox moiety k_A), the concentration of substrate in the solution (C_S), the concentrations of enzyme and redox carrier in the hydrated film (C_E and C_A), the diffusion coefficient of "electrons" and substrate in the film (D_A and D_S), and the thickness of the film (ℓ). Many different situations can be encountered depending on the values of the parameters, and Bartlett and Pratt have identified various distinct cases [54] (Table 2). (See [55] for a more complex case where a redox active inhibitor (O_2) also diffuses in the film and reacts with the enzyme and the mediator.) The reaction

Table 2 A few possible situations encountered when a enzyme is entrapped in a redox hydrogel (Fig. 1g), as described in [54]. The maximal current is that reached on the plateau at high driving force

Case	limitation	depletion	maximal current density	concentration profiles
I	mediator	none	$nFC_Ak_AC_E\ell$	
II	mediator	mediator	$nF\sqrt{D_A}C_A\sqrt{k_AC_E}$	
III	both	both	$nF(D_AC_A + D_SC_S)/\ell$	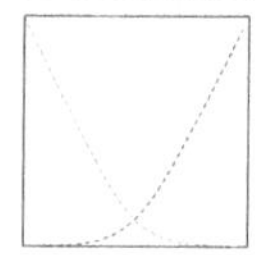
IV	substrate	substrate	$nFC_S\sqrt{k_{\mathrm{cat}}C_ED_S/K_M}$	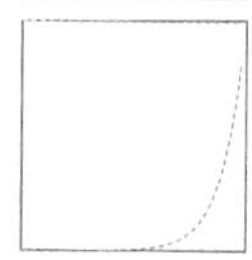
V	substrate	none	$nF\frac{k_{\mathrm{cat}}}{1+\frac{K_M}{C_S^\infty}}C_E\ell$	

of either the enzyme and the mediator, or the enzyme and the substrate, depending on which of k_AC_A and $k_{\mathrm{cat}}/1+\frac{K_M}{C_S^\infty}$ is the smaller, may limit the current. If the reaction is limited by the mediator, for example, there is no dependence of the current on substrate concentration or Michaelis–Menten parameters. If the film is thin enough, both mediator and substrate are approximately homogeneous and catalysis proceeds throughout the film; the current is proportional to the film thickness and the concentration of enzyme and depends on whether the limitation comes from the reaction of the enzyme with the mediator (case I in Bartlett's paper) or the substrate (case V). In case I it is proportional to the concentration of mediator and is independent of substrate concentration whereas in case V its dependence on bulk substrate concentration follows Michaelis–Menten kinetics, irrespective of mediator concentration. For thicker films, catalysis only occurs in a first-order reaction layer (cf. (10)), either at the electrode surface if the process is mediator limited and the concentration of mediator exponentially falls to zero within the film (case II), or at the solution interface if the process is substrate-limited (case IV). In both cases, the current is independent of film thickness, and either proportional to $C_A\sqrt{C_E}$ (case II) or $C_S\sqrt{C_E}$ (case IV). Another case, case III, is observed when both mediators and substrate are consumed within the film, resulting in a change from mediator to substrate-limited kinetics through the film; catalysis only occurs in a thin reaction layer where the concentrations of substrate and mediator (in the right oxidation state) are both non-zero, and the maximal current decreases as the thickness increases, irrespective of enzyme concentration (see [55] for an example). A is the mediator, E the enzyme, S the substrate, ℓ the thickness of the film. The small plots show the mediator (green) and substrate (red) concentration profiles as a function of the distance from the electrode (left) to the solution (right).

15 Considering Adsorbed Enzymes, When, Why, How Fast Should I Rotate the Working Electrode?

If we are considering the situation where the catalyst diffuses to/from the electrode (Fig. 1b), the literature only considers stationary electrodes and there is no rigorous treatment of the reaction/diffusion process at a rotating electrode (see however [56]).

In contrast, considering adsorbed enzymes undergoing direct electron transfer (Fig. 1d), analyzing the signal is only possible if the electrode is rotated to force the convection of the solvent and enhance the transport of substrate towards the electrode. If the electrode is stationary, or if the rotation rate is too slow, mass transport may become the rate-limiting step, in which case the signal becomes entirely independent of the electroactive coverage and the intrinsic properties of the enzyme (Eq. (11) remains valid, but the TOF corresponds to a concentration of substrate that is unknown and possibly much smaller than that in the bulk of the solution, away from the electrode surface). To be able to diagnose this unfavorable situation, one needs to examine whether or not the measured catalytic current is close to the upper limit, i_{Levich}, set by the maximal flux of substrate towards the electrode.

This maximal current is calculated as follows. If the electrode is rotated at an angular velocity ω, it is as if the only effect of the rotation was to set the size of the diffusion layer to a value δ that is proportional to $1/\sqrt{\omega}$:

$$\delta = 1.61 \times D^{1/3}\nu^{1/6}\omega^{-1/2}, \tag{12}$$

where ν is the kinematic viscosity, and δ, D, ν, and ω are in units of cm, cm^2/s, cm^2/s, and rad/s, respectively [1]. The mass transport limited current is reached when the concentration of substrate is zero at the electrode surface, and

$$i_{\text{Levich}} = nFA \times D_s C_s/\delta = 0.62 \times nFA \times D_s^{2/3} C_s \nu^{-1/6}\omega^{1/2}, \tag{13}$$

where C_s is the bulk concentration of substrate, n the number of electrons, and A the electrode area. The concentration of substrate at the electrode $C_s(x=0)$ is given by

$$C_s(x=0) = C_s\left(1 - \frac{i}{i_{\text{Levich}}}\right). \tag{14}$$

Observing that the measured catalytic current is close to i_{Levich} (it cannot be greater) means that the current is controlled by mass transport, and is independent of the electroactive coverage and catalytic properties of the enzyme. In this case, the concentration of substrate at the electrode is much lower than the bulk concentration. In contrast, if the current is much lower than i_{Levich} this means that there is no substrate depletion near the electrode surface. This is the favorable situation. With $D_s = 2 \times 10^{-5}$ cm^2/s, $\nu = 10^{-2}$ cm^2/s (water at 25°C), and $\omega = 500$ rad/s (4,500 rpm), the size of the diffusion layer is about 10 µm, and with $C_s = 1$ mM and $n = 2$, the maximal (mass transport controlled) current density is 4 mA/cm^2. However, if the concentration of substrate is in the µM range, the maximal current density is in the µA/cm^2 range.

Another illustration that mass transport does not limit the current is that the catalytic current is independent of rotation rate. However, it is important to remember that even if the current is limited by mass transport, the dependence of i on ω is weak Eq. (13): increasing the rotation rate fourfold makes the current double at most, and therefore seeing no change in current upon increasing ω from e.g. 5,000 to 6,000 rpm is not conclusive.

Therefore, if the goal is to use direct electrochemistry to study the enzyme's catalytic mechanism, it is more important that the film is stable than the current is high, because large currents have a downside: substrate depletion and mass transport control. To avoid the situation where mass transport is rate limiting, it is important to minimize substrate depletion, even if this requires decreasing the electroactive coverage [46].

16 What Are Steady-State Catalytic Signals and What Can I Learn From Them?

"Steady-state" describes a chronoamperometry experiment where the current is independent of time, or a voltammetric experiment in which the relation between current and potential is independent of scan rate (which usually occurs in the case of a catalytic system if the scan rate is very low). Here we do not count the artifactual decrease in current that results from film loss. Regarding the distinction between catalytic and non-catalytic, steady-state only concerns the former.

Regarding adsorbed enzymes (Fig. 1d), because the non-catalytic signals are most often faint, the electroactive coverage is unknown and the TOF cannot be deduced from the magnitude of the steady-state current (the two are related by Eq. (11)). However, what really matters is not the absolute magnitude of the current, it is its relative change in response to a change in experimental parameters: concentration of substrate, product or inhibitor, light [14] or, most importantly, electrode potential. The change in steady-state catalytic current with substrate or inhibitor concentration should be interpreted just as a change in "initial rate," as described in all enzyme kinetics textbooks [57, 58]. Interpreting the dependence on potential requires that specific kinetic models be used.

As an illustration, Fig. 7 shows the nitrate reduction current obtained with periplasmic nitrate reductase adsorbed at a graphite electrode. Figure 7a, d shows the evolution of nitrate concentration against time, when the concentration is increased stepwise by adding aliquots of a stock solution of potassium nitrate (note the logarithmic *Y* scale). Figure 7b, e shows the resulting change in catalytic current. The steady-state is not the same before and after an injection, but the system quickly relaxes towards a new steady state after each increase in nitrate concentration. Figure 7c, f shows the steady-state catalytic current reached at the end of each step as a function of nitrate concentration. The fit of the Michaelis–Menten equation to the data in C

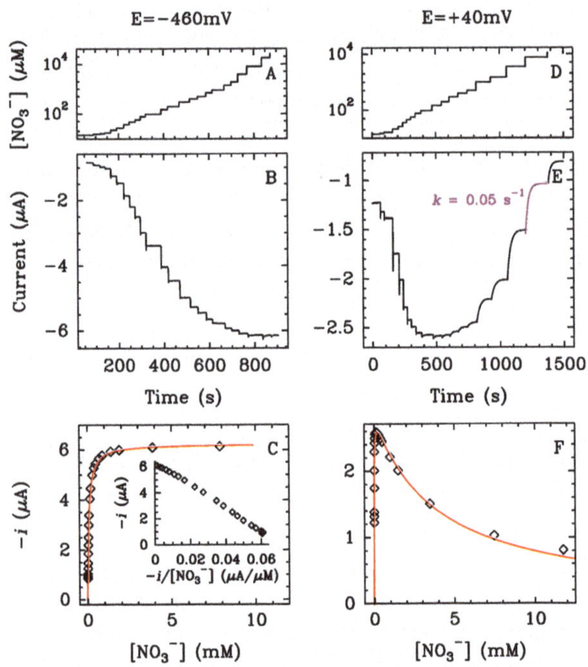

Fig. 7 Dependence of the rate of nitrate reduction by periplasmic nitrate reductase on nitrate concentration. The enzyme is adsorbed onto a rotating graphite electrode. The *left*- and *right*-hand sides correspond to a redox poise at −460 and +40 mV vs SHE, respectively. The data were corrected for film loss as described in [41]. See text. Reprinted with permission from [25]. Copyright (2010) American Chemical Society. Panels (**a**) and (**d**) show nitrate concentration against time, panels (**b**) and (**e**) current against time, and panels (**c**) and (**f**) current against nitrate concentration. The left and right columns show the results of experiments carried out at −460 mV and +40 mV, respectively

$$i_{\text{cat}} = \frac{i_{\max}}{1 + \frac{K_m}{C_S}}, \tag{15}$$

gives $K_m = 85$ μM. The inset shows an Eadie–Hofstee plot. Figure 7e shows that when the same experiment is repeated under more oxidizing conditions, the activity decreases as the concentration of substrate becomes very high (this is called "substrate inhibition"). The red line in Fig. 7f is the best fit of the equation below, with $K_m = 10$ μM and $K_i = 4$ mM.

$$i = \frac{i_{\max}}{1 + \frac{K_m}{C_S} + \frac{C_S}{K_i}}. \tag{16}$$

Each fit also returns a value of $i_{\max}$, which is proportional to the unknown electroactive coverage.

The same kind of experiments can be performed very easily with enzymes that transform or are inhibited by small (gaseous) molecules such as H_2 or CO. In that case, the current can be monitored when the concentration of substrate or inhibitor is changing, e.g., because it is flushed away from the electrochemical cell by bubbling argon. The subtlety is that the current changes as a function of time but steady-state conditions apply at all times, because the concentration of substrate/inhibitor varies much more slowly than the time needed for the system to relax towards a new steady state.

Figure 8 shows data recorded with *Desulfovibrio fructosovorans* NiFe hydrogenase, an enzyme that oxidizes H_2. Figure 8a shows the change in concentration of

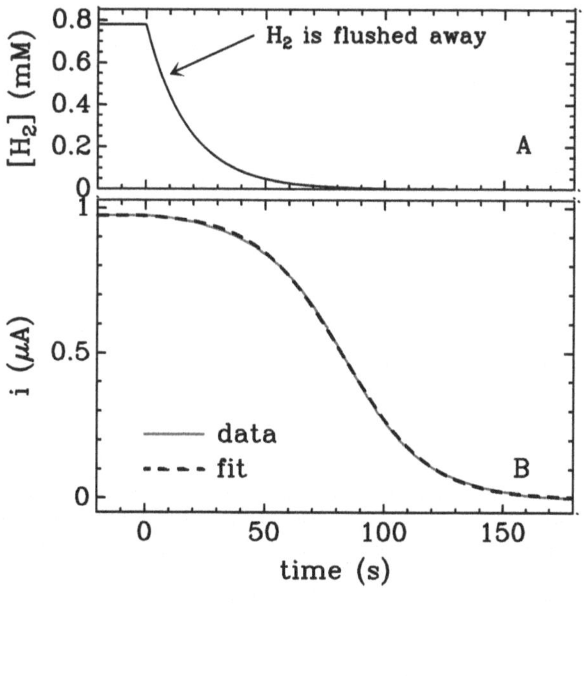

Fig. 8 Hydrogen oxidation by a NiFe hydrogenase: measurement of the Michaelis constant relative to H_2. (**a**) Change in hydrogen concentrations against time. (**b**) Change in current against time in the experiment where H_2 is flushed away at $t > 0$. The fit of (17) is shown as a *dashed line*. Adapted from [4]

dissolved H_2 against time obtained when, having maintained an atmosphere of 1 bar of H_2 until $t = 0$, a small tube is suddenly plunged into the buffer and used to bubble argon. This flushes hydrogen away from the cell and its concentration decreases exponentially with time, with a time constant which we denote by τ_{H_2}. The red curve in Fig. 8b shows the sigmoidal change in current for hydrogen oxidation in the same experiment. The sigmoidal shape of the signal is simply predicted by inserting into the Michaelis–Menten equation a substrate concentration whose time-dependence is $C_{H_2}(t) = C_{H_2}(0)\exp(-t/\tau_{H_2})$:

$$i_{cat} = \frac{i_{max}}{1 + \frac{K_m}{C_{H_2}(0)}\exp(t/\tau_{H_2})}. \tag{17}$$

See Fig. 54 in [4] for a similar experiment where the concentrations of both substrate H_2 and inhibitor CO are simultaneously varied to measure the K_m for H_2 and the K_i for CO in a single experiment, and [59] for experiments aimed at examining the inhibition by H_2 of H_2 production by NiFe and FeFe hydrogenases.

The change in steady-state catalytic current against electrode potential that is recorded in voltammetric experiments (where the electrode potential is swept up and down) gives information which is more original than that discussed above because it often cannot be obtained from solution assays. Figure 9 shows five steady-state voltammograms (Fig. 9a–e) obtained with different enzymes which illustrate the variety of "wave shapes" that can be observed.

Interpreting any of the signals in Fig. 9 requires that at least one kinetic model be used and the corresponding system of equations be solved. In our opinion, the misinterpretation of catalytic voltammograms has two origins. Enzyme kinetics is a quantitative science (it has always been [60, 61]), and *qualitative* interpretations of kinetic data are at best speculative, more often wrong. The second source of mistakes is the same as with the interpretation of any kinetic data, possibly any data: obtaining agreement between a model and an experiment does not imply that the assumptions of the model are correct. For example, it has been known for a century that the Michaelis–Menten steady-state rate equation can be derived by assuming (1) that the enzyme *quickly and reversibly* forms an enzyme substrate (ES) complex that decomposes in a first-order process to give the product or (2) that the ES complex is formed in an *irreversible* second-order reaction, or even (3) that the ES complex is not part of the catalytic cycle. If one forgets that a kinetic model is never unique, one may believe that the agreement between the data and the Michaelis–Menten equation proves that one of these conflicting hypotheses is correct, but it is actually not possible to discriminate between these models based on steady-state measurements alone. Also problematic is the fact that the physical meaning of the parameters in the Michaelis–Menten equation, e.g., K_m, are model dependent (Table 3). This cannot be forgotten when one aims at modeling steady-state kinetic data, whether they come from solution assays or electrochemical experiments. The same "problem" occurs when one considers the interpretation of the "catalytic potential," whose interpretation is also model dependent [3, 18, 20, 60, 61].

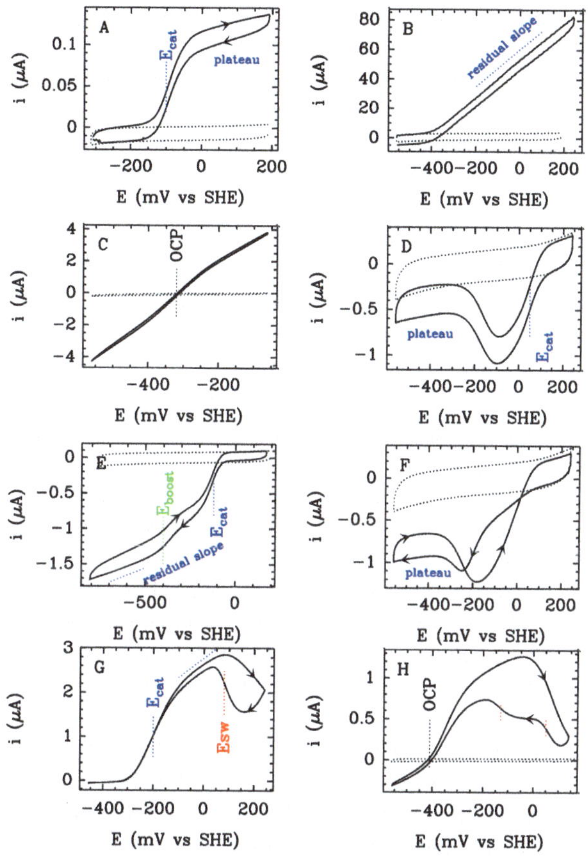

Fig. 9 Catalytic voltammograms obtained with various enzymes adsorbed at a pyrolytic graphite edge rotating electrode. The *dotted lines* are control experiments recorded with no enzyme. OCP is short for "open circuit potential." (**a**) Succinate oxidation by *Escherichia coli* fumarate reductase [20]. (**b**) Hydrogen oxidation by *Allochromatium vinosum* NiFe hydrogenase [21]. (**c**) H^+ reduction and H_2 oxidation by *Chlamydomonas reinhardtii* FeFe hydrogenase [3]. (**d**) Nitrate reduction by *Rhodobacter sphaeroides* nitrate reductase [22]. (**e**) Succinate oxidation by *E. coli* fumarate reductase [23, 24]. (**f**) Nitrate reduction by *R. sphaeroides* nitrate reductase [25]. (**g**) H_2 oxidation by *A. aeolicus* NiFe hydrogenase [26]. (**h**) H^+ reduction and H_2 oxidation by a mutant of *Chlamydomonas reinhardtii* FeFe hydrogenase (unpublished)

Table 3 Illustration that distinct kinetic models give the same rate equation in solution assays, here the Michaelis–Menten equation ($v = v_m/(1 + K_m/C_S)$), but the meaning of the Michaelis parameters v_m and K_M is model dependent

Model	v_m	K_M
$E + S \overset{k_1/k_{-1}}{\rightleftharpoons} ES \overset{k_2}{\rightarrow} E + P$	$k_2 C_E$	$\frac{k_2+k_{-1}}{k_1} > K_d$
$E + S \overset{K_d}{\rightleftharpoons} ES \overset{k_2}{\rightarrow} E + P$	$k_2 C_E$	K_d
$E + S \overset{k_1}{\rightarrow} ES \overset{k_2}{\rightarrow} E + P$	$k_2 C_E$	$\frac{k_2}{k_1}$
$E + S \underset{k_{-1}}{\overset{k_1}{\rightleftharpoons}} ES \overset{k_2}{\rightarrow} ES' \overset{k_3}{\rightarrow} E + P$	$\frac{k_2 k_3 C_E}{k_2+k_3}$	$\frac{k_2}{k_2+k_3} \frac{k_{-1}+k_2}{k_2}$

Interpreting catalytic voltammograms is similar to interpreting initial rates in traditional enzyme kinetics, except that in order to describe a change in activity against potential it is necessary that some of the steps in the kinetic model depend on electrode potential. Over the last 20 years, the wave shapes corresponding to various models have been described, so that it is now possible to draw some general conclusions as to which assumptions lead to which shapes and which interpretation of the signal. Of particular interest is the meaning of the phenomenological "catalytic potential", E_{cat}, defined as the mid-point potential of the sigmoidal catalytic wave, indicated in Fig. 9 by vertical lines.

Because catalysis requires that the enzyme's active site be in a particular redox state, the catalytic wave reports on the redox transformation of the active site. However, just as a Michaelis–Menten constant is not always a dissociation constant (cf. Table 3), the catalytic potential may or may not be a thermodynamic parameter. It is the reduction potential of the active site (at a certain pH and substrate concentration) only if substrate binding and electron transfer between the electrode and the enzyme's active site are reversible and fast compared to the chemical transformation occurring at the active site. (Strictly speaking: compared to the rate of a unique chemical step that determines the rates of catalysis.) This situation is rare and has been documented only a couple of times [19, 20]. Figure 9a shows an example where the enzyme fumarate reductase catalyzes the oxidation of succinate (the reverse of the physiological reaction), and this reaction is so slow that the active site is in Nernstian equilibrium with the electrode; the wave shape simply reports on the potential of the active site flavin, whose dependence on pH and substrate concentration can be used to determine the acidity and dissociation constants that are relevant to the catalytic cycle [20]. However, situations other than that described above give a catalytic potential the interpretation of which is entirely different. In particular, the effects of slow substrate binding [22, 62] and slow electron transfer [3, 63] have been discussed in relation to the meaning of E_{cat}.

In Fig. 9a the current tends to a limit at high potential (a plateau) because under these conditions the active site is fully oxidized in the steady state, and further increasing the potential makes no difference. The limiting value of the current is proportional to the TOF of the fully oxidized enzyme. The situation is different in Fig. 9e, which shows fumarate reduction by fumarate reductase, and the current is a linear function of E in the low potential region, where we would expect to see a plateau (see also Fig. 9b, c). The explanation for this is that the rate of interfacial ET (between the electrode and the surface exposed redox relay) is not very fast compared to turnover, and therefore it partly determines the turnover rate. Because all the enzyme molecules cannot be adsorbed in exactly the same orientations, there must be distribution of distances between the electrode and the enzyme (or the surface exposed redox center, when an electron transfer chain wires the active site to the surface), and therefore a distribution of rates of interfacial ET, and a distribution of turnover rates [19]. At a given potential, the enzyme molecules that contribute to the catalytic current are those for which the interfacial electron transfer rate are comparable to the rate of catalysis or greater. As the driving force increases, the interfacial electron transfer rates increases, cf. Eq. (9), so that enzymes with smaller k_0 contribute to the catalytic current. If there is no plateau, the limiting slope (not the current) is proportional to the TOF of the fully oxidized enzyme (Eq. (10) in [19] or Eq. (37) in [4]).

Simple wave shapes such as in Fig. 9a can also occur if one intramolecular ET step, along the chain that wires the surface exposed relay to the active site, is slow. Under special circumstances (irreversible catalysis and conditions defined in the discussion of Eq. (5) in [63]) the catalytic potential may equate the standard potential of one of the relays [47], but in the general case it has no simple meaning (cf. the discussion of Eq. (12) and Eq. (13) in [3]). There is no basis to the claim that the catalytic potential necessarily is "the reduction potential of the electron relay center at which electrons enter or leave the enzyme" – it is so in just one very particular case.[1]

Sometimes the steady-state signals have very complex shapes, as illustrated in Fig. 9d, e by fumarate reductase and periplasmic nitrate reductase. The catalytic current is zero at high potential because the enzyme is oxidized; activity kicks as the potential is lowered and increases in a sigmoidal manner; however, the activity drops at lower potential in the case of nitrate reductase and other, totally unrelated enzymes (cytochrome c nitrite reductase [51, 65–67] and succinate dehydrogenase [68]) whereas in the case of fumarate reductase the wave shape shows a boost in activity at low potential (Fig. 9e). Two distinct explanations have been offered. One is that at a certain potential, the redox state of a redox relay remote from the active site changes, and this either changes the rate of electron transfer along the chain or affects the property of the distant active site by an allosteric effect. There is indeed experimental evidence that the catalytic boost of fumarate reductase occurs at the standard potential of an iron sulfur cluster than is in the middle of the electron transfer chain [24], and that the rate of release of a particular inhibitor (oxaloacetate) from the active site depends on the redox state of this remote cluster [23]. Another explanation is that because changing the potential changes the rates of the redox steps in the catalytic cycle, it may be that catalysis takes a different route depending on potential. The latter hypothesis has been turned into kinetic models that explain the wave shape of nitrate reductase [22, 62, 69], but there is still much to understand about these complex wave shapes.

Note that if the enzyme can perform oxidative and reductive catalysis, as observed in Fig. 9c with a hydrogenase that oxidizes H_2 under oxidizing conditions and reduces protons at low potential, then the current is zero at a particular potential, called the open circuit potential (OCP) which equates the equilibrium potential of the substrate/product redox couple [20, 46]. Measuring this potential

---

[1] This claim is based on a model that considers this electron relay center as the only redox center in the enzyme [64], which, of course, can only predict that the catalytic potential is the potential of this center. Models that take into account the redox transformations of the active site and of an electron transfer relay (and electron transfer between the two) predict that the catalytic potential may or may not match the potential of one of these redox centers, depending on how the rates of intramolecular electron transfer compare to those that describe the transformations of the active site [3, 63].

tells us nothing about the enzyme (except that the enzyme works in both directions) but it can be valuable information if the two species, substrate and product, do not interconvert in the absence of a catalyst, which makes the equilibrium potential of the substrate/product redox couple impossible to measure in titration experiments (see [20, 70, 71] for examples).

17 Regarding Catalytic Signals, What Is a Hysteresis and What Does It Mean?

Hysteresis is a departure from steady-state, which occurs when the current response depends not only on potential but also on time or scan rate, and on the previous history of the experiment. Hysteresis may either be caused by the diffusive transport of redox species towards the electrode (if the latter is not rotated or not rotated fast enough), or reflect the evolution over time of the properties of the catalyst.

The textbook case of hysteresis in physics is the magnetization response of a ferromagnet such as iron in response to a magnetic field. The typical example in protein electrochemistry is the activity of hydrogenase: these enzymes slowly and reversibly convert in an inactive form under very oxidizing conditions. In voltammetric experiments where the enzyme is adsorbed onto an electrode this is observed as a decrease in catalytic current as the potential increases and a recovery of activity on the sweep to low potential, but the upward and downward current responses do not overlap. The hysteresis disappears at fast scan rates, when the time needed to record the CV is smaller than the time needed for the enzyme to inactivate (see, e.g., the purple line in Fig. 10a). The hysteresis should also disappear when the scan rate is so slow that the (in)activation remains at equilibrium on the voltammetric time scale (this situation has never been observed).

On the sweep towards low potential, the reactivation is sudden and the enzyme "switches back on" at a potential that is usually called "E_{sw}", shown as vertical lines in Figs. 9g and 10a. To illustrate the semi-quantitative interpretation of complex catalytic signals, we explain below that this switch potential can be given a simple meaning. (Note that the wave shapes in Figs. 9g and 10a can be simulated using numerical [26, 75, 81] or analytical [3, 72] models.) We describe the activation/inactivation by the following reaction:

$$\mathrm{A} \underset{k_a}{\overset{k_i}{\rightleftharpoons}} \mathrm{I}, \tag{18}$$

where A and I are the active and inactive forms of the enzyme, which interconvert with potential-dependent, first-order rate constants k_i and k_a. At any time and potential, the current is proportional to the total electroactive coverage, the time-dependent fraction of enzyme that is in the active form $x_A = [\mathrm{A}]/([\mathrm{A}]+[\mathrm{I}])$, and the potential-dependent steady-state turnover rate of the active enzyme (again we

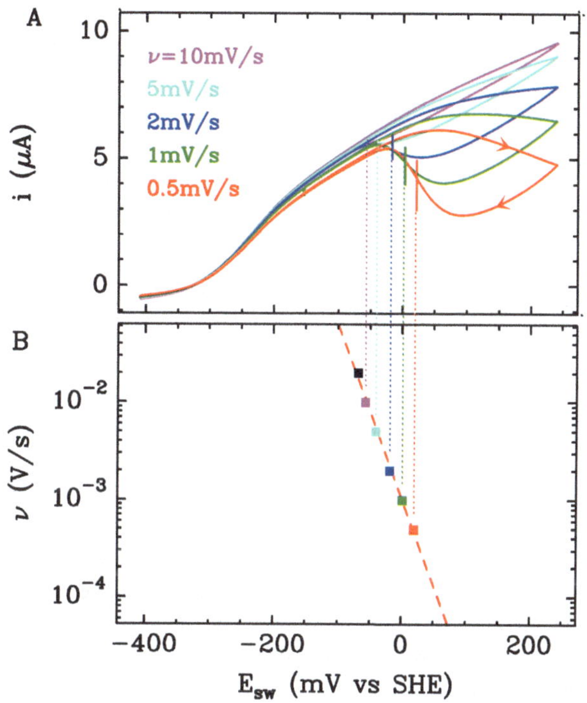

Fig. 10 How the "switch potential" of NiFe hydrogenase depends on scan rate. (**a**) Series of catalytic voltammograms recorded at different scan rates, with a mutant of *D. fructosovorans* NiFe hydrogenase adsorbed at a rotating graphite electrode. (**b**) *Straight line* fitted to the change in E_{sw} against log of scan rate. Adapted from [72]

assume that the fraction of active enzyme varies much slowly than the time needed for the catalytic cycle to relax towards a new steady-state at each potential). The shape of the catalytic signal on the scan to low potential can be understood by taking into account that at low enough potential, k_a is greater than k_i, and increases exponentially as the potential decreases:

$$k_a(E) = k\exp\left[-\frac{F}{RT}E\right]. \tag{19}$$

This equation applies for all (NiFe, NiFeSe and FeFe) hydrogenases [26, 72–74]. This exponential dependence of k_a on E is the reason the increase in activity is very sharp during the sweep towards low potential. A simple dimensional analysis predicts the relation between k_a and the switch potential. Reactivation occurs when E is so low that the time scale of reactivation, $1/k_a$, becomes similar to the voltammetric time scale:

$$k_a(E = E_{\text{sw}}) = F\nu/RT. \tag{20}$$

Equation (20) gives the rate of reactivation at $E = E_{\text{sw}}$. See SI text in [72] for a rigorous demonstration. In [73], Eq. (20) was used to determine the dependence of k_a on E from a series of voltammograms recorded at various scan rates. Combining Eq. (19) and Eq. (20) gives the relation between E_{sw} and the scan rate:

$$E_{sw} = \frac{RT}{F}\log\frac{F}{RTk} - \frac{RT}{F}\log(v). \tag{21}$$

Such linear relation between E_{sw} and log of scan rate has been observed in all voltammetric experiments with hydrogenase where the scan rate was varied (see Fig. 10, the experiments with *D. fructosovorans* NiFe hydrogenase in [72] and Fig. S4 in [26]; *Aquifex aeolicus* NiFe hydrogenase, Fig. 6 in [26]; *Desulfovibrio vulgaris* MF NiFe hydrogenase, Fig. S4 in [76]; *D. vulgaris* NiFeSe hydrogenase in [73]); it is important because it tells us that the switch potential is not defined by the standard potential of the inactive state (which is not a parameter in this equation), but rather by the kinetics of reactivation.

Hydrogenases were known to inactivate under oxidizing conditions well before they were studied using electrochemistry, but electrochemical experiments have shown that other enzymes slowly interconvert between active and less active (or inactive) forms as the potential is varied. An example is shown in Fig. 9f with the same nitrate reductase as in Fig 9d, but under slightly different experimental conditions (higher nitrate concentration). A very strong hysteresis is observed, which reveals that the enzyme inactivates under oxidizing conditions and reactivates at lower potential, and these reactions are slow on the time scale of the voltammetry. These signals can be simulated by combining the information about the kinetics of (in)activation and the complex dependence of a steady-state TOF on potential [75].

18 When Should I Use Chronoamperometry and Why?

Chronoamperometric measurements of the TOF of enzymes directly connected to electrodes is particularly useful in two different contexts: the enzyme slowly (in) activates in response to a change in electrode potential or one wants to follow the change in activity that results from a change in substrate or inhibitor concentration.

When the goal is to study the redox-driven (in)activation of an enzyme, using chromoamperometry experiments is often simpler than voltammetry, because the kinetics of (in)activation is detected without the convolution with the dependence of activity on potential that gives voltammograms their very complex shapes. Current transients that follow potential steps are often exponential (see however [74, 76]), and both the magnitude and the time constant of these exponential transients may embed meaningful information.

For example, Fig. 11 shows how the activity of periplasmic nitrate reductase varies after stepping the electrode potential when the nitrate concentration is high and the CV looks similar to that in Fig. 9f. After each step, the activity relaxes exponentially toward a new steady-state value; it slowly increases (the current becomes more negative) after a step down and slowly decreases after a step up. Both the asymptotic value of the current and the rate of relaxation depend on potential, but neither depends on the direction of the step or sample history. This is

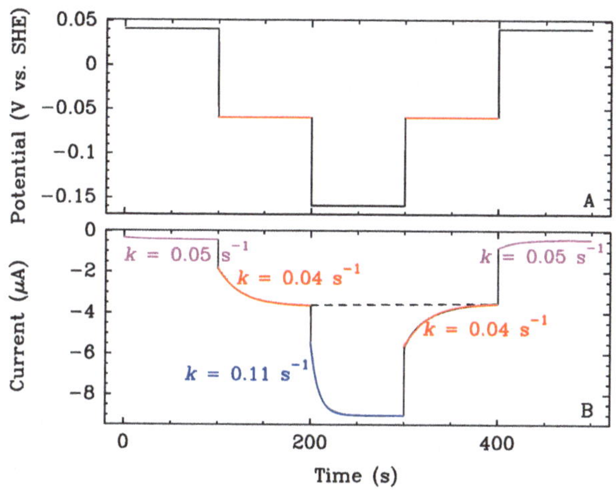

Fig. 11 Redox-driven reversible (in)activation of periplasmic nitrate reductase adsorbed at a rotating graphite electrode. The turnover rate relaxes towards a new steady state after stepping the electrode potential. (**a**) Potential that was applied. (**b**) Resulting current. The rate constants k determined by fitting each transient to an exponential function are indicated for each step. The data were corrected for film loss as described in [41]. Reprinted with permission from [25]. Copyright (2010) American Chemical Society

understood by considering that the active and inactive forms of the enzyme, A and I, respectively, interconvert with potential-dependent rate constants, as in (18). The steady-state fraction of inactive enzyme at any potential depends on the equilibrium constant k_i/k_a at that potential. This implies that the asymptotic value of the current depends only on potential, but the magnitude of the transient depends on the history of the sample (compare the two red traces recorded at the same potential in Fig. 11). The rates k of the exponential relaxations in Fig. 11 are neither k_a nor k_i, but rather the sum of the two, $k = k_i + k_a$.[2]

Note that at −60 mV, for example, the rate constant of the relaxations (red lines) is independent of whether the enzyme is activating ($t = 100$–200 s) or inactivating (300–400 s), and the asymptotic value of the current at $t = 200$ and 400 s is also the same (this is marked by the horizontal dashed line). The relaxation plotted in purple corresponds to the same conditions as those of the purple transient in Fig. 7e. This is because the rate of relaxation towards steady state is the same whether the transformation between active and inactive forms is triggered by changing the potential or the nitrate concentration. This method has also been used repeatedly to study the reversible (in)activation of hydrogenases [26, 73, 74, 77].

[2] To demonstrate this, let us consider the transformation between A and B according to $A \underset{k_{-1}}{\overset{k_1}{\rightleftharpoons}} B$ The change in concentrations obeys the differential equation $d[A]/dt = -k_1[A] + k_{-1}(C_0 - [A])$, where $C_0 = [A] + [B]$. The solution is $[\mathrm{A}] = D \times \exp[-(k_1 + k_{-1})t] + C_0 k_{-1}/(k_1 + k_{-1})$ where D is an integration constant.

Chronoamperometry can also be used to measure the magnitudes and rates of irreversible transformations of an enzyme. By "irreversible" here we mean that, under particular experimental conditions, the system tends towards a state that is either fully active or fully inactive, and the time constant of the transformation is either $1/k_i$ or $1/k_a$, which can be measured from the time constant of the exponential change in current in chronoamperometry experiments. In that case, the magnitude of the transient is directly proportional to the amount of enzyme that converts into an active or inactive form.

For example, periplasmic nitrate reductase irreversibly activates the first time it is reduced and the magnitude of the activation correlates with the magnitude of a certain EPR signature in the sample before activation, from which we concluded that this signature is that of an inactive state [11]. Membrane-bound nitrate reductases also irreversibly activate upon reduction [76, 78], but the process is biphasic and could not be related to the disappearance of EPR-active species [76]. NiFe hydrogenases inactivate under oxidizing conditions (either in the presence or in the absence of O_2) to form a mixture of inactive states; irreversible reactivation can be triggered by stepping the potential down, and the biphasic increase in current can be used to measure the amounts of the different inactive states [72, 79, 80].

Finally, chronoamperometry can be very useful to measure the change in steady-state turnover rate when the concentration of substrate or inhibitor varies (we have seen examples of this above), or the rates of inhibitor binding or release as described below (Fig. 12). The experiment is very easy with gaseous inhibitor

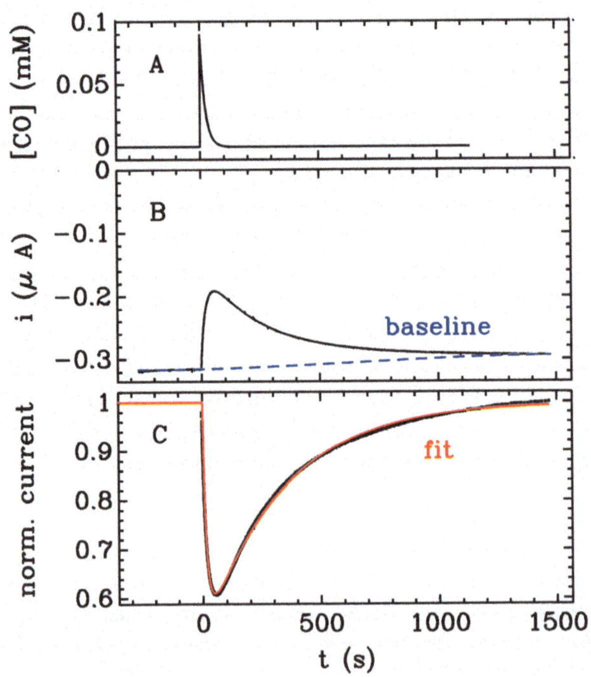

Fig. 12 Experiment showing the effect of a transient exposure to the inhibitor CO on the rate of nitrite reduction by cytochrome *c* nitrite reductase adsorbed at a pyrolytic graphite edge rotating electrode. (**a**) Concentration of CO against time. (**b**) Catalytic current (negative for a reduction), and the baseline as a *dashed line*. (**c**) Current corrected for film loss by dividing the signal by the baseline, and the fit of (2) in [51]. Adapted from Fig. S1 in [51]

because they can be flushed away from the cell solution by bubbling a neutral gas such as argon, resulting in an exponential decay of the concentration [53, 81] (Fig. 12a).

This is useful when the reaction with the inhibitor is irreversible (e.g., NiFe hydrogenase exposed to O_2 [53], FeFe hydrogenase exposed to CO under reducing conditions [82]) or reversible (e.g., NiFe hydrogenase or cytochrome c nitrite reductase exposed to CO [51, 83]). Figure 12 shows an example where the negative current that is proportional to the rate of nitrite reduction by cytochrome c nitrite reductase changes when the film is transiently exposed to CO. CO is added in the electrochemical cell by injecting a small amount of solution saturated with CO, and flushed away by bubbling argon. The concentration of CO decreases with a time constant of about 20s. The change in activity against time can be analyzed to measure the rates of CO binding and release (Fig. 12) [51]. Based on similar experiments carried out with NiFe hydrogenases, we could show that the rate of CO binding and release informs on the rate of diffusion along the gas channel that connects the active site to the solvent [83–85], which gives information about intramolecular diffusion rates which is impossible to obtain using conventional techniques. This is an interesting example where direct electrochemistry is used to learn about a non-redox step in the catalytic mechanism.

19 How Can I Detect and Correct Film Loss?

For catalytic electrochemistry with adsorbed enzymes, the signal is always proportional to the total electroactive coverage, the fraction of enzyme that is active (in case there is a transformation between active and inactive forms), and the TOF of the active enzyme. Therefore, the decrease in electroactive coverage over time makes the current decrease. This effect must be corrected to obtain a signal that reflects the behavior of the enzyme.

Note that film loss and irreversible inactivation have exactly the same effect, proving that a decrease in current results from inactivation requires that one is able to find conditions where the initial current is recovered. If the decrease in activity cannot be reversed, it is very difficult to discriminate between film loss and irreversible inactivation (see however the discussion of Fig. S4 in [74]).

Correcting for film loss (or irreversible inactivation) is rather easy on condition that film loss is independent of any other process that makes the current change. The current is the product of independent contributions: the time-dependent electroactive coverage $\Gamma(t)$, whose influence we seek to remove, and the rate of catalysis $k(t)$, which is the signal we are ultimately interested in. Both contributions (coverage and activity) can depend on time, electrode potential, and substrate or inhibitor concentration.

$$i_{\mathrm{cat}} = nFA\Gamma(t) \times x_{\mathrm{A}} \times \mathrm{TOF} = \Gamma(t) \times k(t). \tag{22}$$

In that case, there surely is a way to record independently or calculate a control signal, $i'(t)$, which is proportional to the time-dependent electroactive coverage during the principal experiment:

$$i' = \alpha \times \Gamma(t). \tag{23}$$

Dividing $i_{\mathrm{cat}}(t)$ by the control signal $i'(t)$ yields the signal of interest, $k(t)$, up to a factor of proportionality $1/\alpha$. That this factor is unknown is not a problem because in experiments with adsorbed enzymes the current is proportional to the unknown electroactive coverage. Several examples of how this strategy can be used are illustrated in [41].

As an example, Fig. 12b shows the baseline calculated from the initial and final parts of the chronoamperogram, and the normalized signal in Fig. 12c was obtained by dividing the current by this baseline.

This method also applies to experiments where electron transfer is mediated, rather than direct, providing the current is proportional to the time-dependent concentration of catalyst.

20 Does the Electrode Affect the Properties of the Protein/ Enzyme?

This is a very good question, which must be asked if the protein of interest directly exchanges electrons with the electrode, whether it is adsorbed or diffuses in solution. This question has no general answer because there are many examples where the protein clearly denatures on the electrode surface, and many others where the TOF is actually greater when the enzyme is adsorbed than when it is in solution (this can happen if electron exchange with the soluble redox partner is slower than with the electrode; see, e.g., [46] and Note 34 in [11]). So the question should be considered on a case-by-case basis.

There are in the literature a number of examples where it is clear that the electrochemical signal obtained with an adsorbed enzyme has nothing to do with the properties of the system as observed in solution titrations or activity assays. Sometimes a non-catalytic peak is observed at a potential shifted more than 100 mV from the value expected from solution titration. Or the non-catalytic signal consists of several peaks which should have the same intensity, but they don't. Or the value of k_0 determined by examining how the signal depends on scan rate is very low ($<1\ \mathrm{s}^{-1}$ or so). Sometimes the onset of the catalytic current occurs at much lower or greater potential than expected based on what we know about the potential of the active site catalytic intermediates. Sometimes a catalytic current is observed for a substrate that is not the physiological substrate, or the magnitude of the catalytic current is low despite the fact that the electroactive coverage is high, or a catalytic current is observed with a mutant that is completely inactive in solution assays, or the catalytic current is not affected by a molecule which is known to inhibit the activity in solution assays. Spectroscopic experiments with proteins or enzymes adsorbed onto electrodes have shown structural modifications that occur upon adsorption [86, 87]. Any of these observations should be considered a red flag.

However, the point of using electrochemistry is to make observations that would escape detection using traditional methods, so how can we make sure that these are not artifacts? It is not always possible, but it is at least fair to try to design solution experiments that support the conclusions from electrochemical experiments. For example, the increase in activity as the driving force *decreases*, observed with nitrate reductase (Fig. 9d) and succinate dehydrogenase, could be mirrored in solution experiments where the redox partner is the limiting reactant and the activity is increases as the redox partner is consumed and the driving force provided by the redox partner decreases [88, 89]. The potential-dependent substrate inhibition seen in Fig. 7 was backed up by the results of solution assays with redox partners having different potentials [25]. Regarding NiFe hydrogenase, the effect of mutations on the rate of CO diffusion measured by electrochemistry could be compared to the results obtained in a completely different method based on a solution assay [83, 90]. The conclusion based on electrochemistry experiments that two inactive states of NiFe hydrogenase may be formed under anaerobic oxidizing conditions was corroborated by experiments where the enzyme was chemically oxidized and examined by EPR [80]. In all these cases the solution experiments cannot reach the same level of detail or accuracy, but they may give credence to the electrochemical experiments.

21 Can I Read More About All This?

Several researchers involved in electrochemical studies of proteins have written reviews; see, e.g., [4, 34, 91–98]. There is no textbook on bioelectrochemistry, but it is always very important to know the general principles of electrochemistry, which have been covered in many very good books; see, e.g., [1, 15, 99]. If you work with enzymes, you have to know about enzyme kinetics, and there are very comprehensive enzyme kinetics textbooks (see, e.g., [57, 58]).

22 Do You Have Any Final Advice?

Two pieces of advice actually: read the literature and run control experiments!

Repeat your experiment with no enzyme to make sure the current you are looking at is not caused by direct reaction of the substrate on the electrode. Repeat the inhibition experiment with no inhibitor to make sure that the inactivation that you are observing is not caused by the electrode potential you are using. Repeat the experiment with hemin or free flavin if you are studying a cytochrome- or FAD-dependent enzyme. Make sure a known inhibitor of the enzyme you're studying actually inhibits the catalytic current that you are recording. Sometimes we go down the wrong road; it is better to figure this out before wasting too much time.

References

1. Bard AJ, Faulkner LR (2004) Electrochemical methods. Fundamental and applications, 3rd edn. Wiley, New York
2. Bard AJ, Inzelt G, Scholz F (2012) Electrochemical dictionary. Springer, Berlin. doi:10.1007/978-3-642-29551-5. ISBN 978-3-642-29550-8
3. Fourmond V, Baffert C, Sybirna K, Lautier T, Abou Hamdan A, Dementin S, Soucaille P, Meynial-Salles I, Bottin H, Léger C (2013) J Am Chem Soc 135:3926–3938
4. Léger C, Bertrand P (2008) Chem Rev 108:2379–2438
5. Wait AF, Parkin A, Morley GM, dos Santos L, Armstrong FAJ (2010) J Phys Chem C 114:12003–12009
6. Lalaoui N, de Poulpiquet A, Haddad R, Le Goff A, Holzinger M, Gounel S, Mermoux M, Infossi P, Mano N, Lojou E, Cosnier S (2015) Chem Commun 51:7447–7450
7. Zhang W, Li G (2004) Anal Sci 20:603–609
8. Pinczewska A, Sosna M, Bloodworth S, Kilburn JD, Bartlett PN (2012) J Am Chem Soc 134:18022–18033
9. Elliott SJ, Hoke KR, Heffron K, Palak M, Rothery RA, Weiner JH, Armstrong FA (2004) Biochemistry 43:799–807
10. Pershad HR, Hirst J, Cochran B, Ackrell BAC, Armstrong FA (1999) Biochim Biophys Acta 1412:262–272
11. Fourmond V, Burlat B, Dementin S, Arnoux P, Sabaty M, Boiry S, Guigliarelli B, Bertrand P, Pignol D, Léger C (2008) J Phys Chem B 112:15478–15486
12. Silveira CM, Almeida MG (2013) Trends Anal Bioanal Chem 405:3619–3635
13. Limoges B, Moiroux J, Savéant J-MJ (2002) J Electroanal Chem 521:8–15
14. Plumeré N (2016) Biophotoelectrochemistry: from bioelectrochemistry to photosynthesis. Adv Biochem Eng. Springer, Chapter 4
15. Saveant JM (2006) Elements of molecular and biomolecular electrochemistry. Willey, Hoboken
16. Jeuken LJC (2016) Biophotoelectrochemistry: from bioelectrochemistry to photosynthesis. Adv Biochem Eng. Springer, Chapter 2
17. Jeuken LJC, Connell SD, Henderson PJF, Gennis RB, Evans SD, Bushby RJ (2006) J Am Chem Soc 128:1711–1716
18. McMillan DG, Marritt SJ, Firer-Sherwood MA, Shi L, Richardson DJ, Evans SD, Elliott SJ, Butt JN, Jeuken LJ (2013) J Am Chem Soc 135:10550–10556
19. Léger C, Jones AK, Albracht SPJ, Armstrong FA (2002) J Phys Chem B 106:13058–13063
20. Léger C, Heffron K, Pershad HR, Maklashina E, Luna-Chavez C, Cecchini G, Ackrell BAC, Armstrong FA (2001) Biochemistry 40:11234–11245
21. Léger C, Jones AK, Roseboom W, Albracht SPJ, Armstrong FA (2002) Biochemistry 41:15736–15746
22. Bertrand P, Frangioni B, Dementin S, Sabaty M, Arnoux P, Guigliarelli B, Pignol D, Léger C (2007) J Phys Chem B 111:10300–10311
23. Heering HA, Weiner JH, Armstrong FA (1997) J Am Chem Soc 119:11628–11638

24. Hudson JM, Heffron K, Kotlyar V, Sher Y, Maklashina E, Cecchini G, Armstrong FA (2005) J Am Chem Soc 127:6977–6989
25. Fourmond V, Sabaty M, Arnoux P, Bertrand P, Pignol D, Léger C (2010) J Phys Chem B 114:3341–3347
26. Fourmond V, Infossi P, Giudici-Orticoni M-T, Bertrand P, Léger C (2010) J Am Chem Soc 132:4848–4857
27. Armstrong FA, Wilson GS (2000) Electrochim Acta 45:2623–2645
28. Morris GP, Baker RE, Gillow K, Davis JJ, Gavaghan DJ, Bond AM (2015) Langmuir 31:4996–5004
29. Salverda JM, Patil AV, Mizzon G, Kuznetsova S, Zauner G, Akkilic N, Canters GW, Davis JJ, Heering HA, Aartsma TJ (2010) Angew Chem Int Ed 49:5776–5779
30. Kemper MA, Gao-Sheridan HS, Shen B, Duff JLC, Tilley GJ, Armstrong FA, Burgess BK (1998) Biochemistry 37:12829–12837
31. Battistuzzi G, Borsari M, Sola M, Francia F (1997) Biochemistry 36:16247
32. Hirst J, Duff JLC, Jameson GNL, Kemper MA, Burgess BK, Armstrong FA (1998) J Am Chem Soc 120:7085–7094
33. Chen K, Hirst J, Camba R, Bonagura CA, Stout CD, Burgess BK, Armstrong FA (2000) Nature 405:814–817
34. Crichton RR, Louro RO (2013) Practical approaches to biological inorganic chemistry, 1st edn. Elsevier, Oxford
35. Clarke TA et al (2011) Proc Natl Acad Sci 108:9384–9389
36. Bewley KD, Ellis KE, Firer-Sherwood MA, Elliott SJ (1827) Biochim Biophys Acta 2013:938–948
37. Chobot SE, Hernandez HH, Drennan CL, Elliott SJ (2007) Angew Chem Int Ed 46:4145–4147
38. Maiocco SJ, Grove TL, Booker SJ, Elliott SJ (2015) J Am Chem Soc 137:8664–8667
39. Zu Y, Fee J, Hirst J (2001) J Am Chem Soc 123:9906–9907
40. Jeuken LJC, van Vliet P, Verbeet MP, Camba R, McEvoy JP, Armstrong FA, Canters GW (2000) J Am Chem Soc 122:12186–12194
41. Fourmond V, Lautier T, Baffert C, Leroux F, Liebgott P-P, Dementin S, Rousset M, Arnoux P, Pignol D, Meynial-Salles I, Soucaille P, Bertrand P, Léger C (2009) Anal Chem 81:2962–2968
42. Jeuken LJC, Armstrong FA (2001) J Phys Chem B 105:5271–5282
43. Baymann F, Barlow NL, Aubert C, Schoepp-Cothenet B, Leroy G, Armstrong FA (2003) FEBS Lett 539:91–94
44. Chidsey CED (1991) Science 251:919–922
45. Turner KL, Doherty MK, Heering HA, Armstrong FA, Reid GA, Chapman SK (1999) Biochemistry 38:3302–3309
46. Pershad HR, Duff JLC, Heering HA, Duin EC, Albracht SPJ, Armstrong FA (1999) Biochemistry 38:8992–8999
47. Elliott SJ, McElhaney AE, Feng C, Enemark JH, Armstrong FA (2002) J Am Chem Soc 124:11612–11613
48. Jones AK, Camba R, Reid GA, Chapman SK, Armstrong FA (2000) J Am Chem Soc 122:6494–6495
49. Aguey-Zinsou KF, Bernhardt P, McEwan A, Ridge J (2002) J Biol Inorg Chem 7:879–883
50. Hoke KR, Cobb N, Armstrong FA, Hille R (2004) Biochemistry 43:1667–1674
51. Almeida MG, Guigliarelli B, Bertrand P, Moura JJG, Moura I, Léger C (2007) FEBS Lett 581:284–288
52. Judd ET, Stein N, Pacheco AA, Elliott SJ (2014) Biochemistry 53:5638–5646
53. Léger C, Dementin S, Bertrand P, Rousset M, Guigliarelli B (2004) J Am Chem Soc 126:12162–12172
54. Bartlett PN, Pratt KFE (1995) J Electroanal Chem 397:61–78
55. Fourmond V, Stapf S, Li H, Buesen D, Birrell J, Rüdiger O, Lubitz W, Schuhmann W, Plumeré N, Léger C (2015) J Am Chem Soc 137:5494–5505
56. Costentin C, Dridi H, Savéant J-M (2015) J Am Chem Soc 137:13535–13544

57. Cornish-Bowden A (2004) Fundamental of enzyme kinetics. Portland Press, London
58. Fersht A (1999) Structure and mechanism in protein science: a guide to enzyme catalysis and protein folding. W H Freeman and Company, New York. ISBN 0716732688 9780716732686
59. Fourmond V, Baffert C, Sybirna K, Dementin S, Abou-Hamdan A, Meynial-Salles I, Soucaille P, Bottin H, Leger C (2013) Chem Commun 49:6840–6842
60. Cornish-Bowden A (2013) FEBS Lett 587:2725–2730
61. Johnson KA (2013) FEBS Lett 587:2753–2766
62. Frangioni B, Arnoux P, Sabaty M, Pignol D, Bertrand P, Guigliarelli B, Léger C (2004) J Am Chem Soc 126:1328–1329
63. Léger C, Lederer F, Guigliarelli B, Bertrand P (2006) J Am Chem Soc 128:180–187
64. Hexter SV, Grey F, Happe T, Climent V, Armstrong FA (2012) Proc Natl Acad Sci U S A 109:11516–11521
65. Gwyer JD, Zhang J, Butt JN, Ulstrup J (2006) Biophys J 91:3897–2906
66. Judd ET, Youngblut M, Pacheco AA, Elliott SJ (2012) Biochemistry 51:10175–10185
67. Lockwood CWJ, Burlat B, Cheesman MR, Kern M, Simon J, Clarke TA, Richardson DJ, Butt JN (2015) J Am Chem Soc 137:3059–3068
68. Sucheta A, Ackrell BAC, Cochran B, Armstrong FA (1992) Nature 356:361–362
69. Heffron K, Léger C, Rothery RA, Weiner JH, Armstrong FA (2001) Biochemistry 40:3117–3126
70. Kurth JM, Dahl C, Butt JN (2015) J Am Chem Soc 137:13232–13235
71. Reda T, Plugge CM, Abram NJ, Hirst J (2008) Proc Natl Acad Sci U S A 105:10654–10658
72. Hamdan AA, Liebgott P-P, Fourmond V, Gutiérrez-Sanz O, De Lacey AL, Infossi P, Rousset M, Dementin S, Léger C (2012) Proc Natl Acad Sci U S A 109:19916–19921
73. Ceccaldi P, Marques MC, Fourmond V, Pereira IC, Leger C (2015) Chem Commun 51:14223–14226
74. Fourmond V, Greco C, Sybirna K, Baffert C, Wang P-HH, Ezanno P, Montefiori M, Bruschi M, Meynial-Salles I, Soucaille P, Blumberger J, Bottin H, De Gioia L, Léger C (2014) Nat Chem 6:336–342
75. Jacques JGJ, Burlat B, Arnoux P, Sabaty M, Guigliarelli B, Léger C, Pignol D, Fourmond V (2014) Biochim Biophys Acta 1837:1801–1809
76. Ceccaldi P, Rendon J, Léger C, Toci R, Guigliarelli B, Magalon A, Grimaldi S, Fourmond V (2015) Biochim Biophys Acta Bioenergetics 1847:1055–1063
77. Jones AK, Lamle SE, Pershad HR, Vincent KA, Albracht SPJ, Armstrong FA (2003) J Am Chem Soc 125:8505–8514
78. Field SJ, Thornton NP, Anderson LJ, Gates AJ, Reilly A, Jepson BJN, Richardson DJ, George SJ, Cheesmana MR, Butt JN (2005) Dalton Trans 3580–3586
79. Lamle SL, Albracht SPJ, Armstrong FA (2004) J Am Chem Soc 126:14899–14909
80. Abou Hamdan A, Burlat B, Gutierrez-Sanz O, Liebgott P, Baffert C, de Lacey A, Rousset M, Guigliarelli B, Leger C, Dementin S (2013) Nat Chem Biol 9:15–17
81. Orain C, Saujet L, Gauquelin C, Soucaille P, Meynial-Salles I, Baffert C, Fourmond V, Bottin H, Léger C (2015) J Am Chem Soc 137:12580–12587
82. Baffert C, Bertini L, Lautier T, Greco C, Sybirna K, Ezanno P, Etienne E, Soucaille P, Bertrand P, Bottin H, Meynial-Salles I, De Gioia L, Léger C (2011) J Am Chem Soc 133:2096–2099
83. Leroux F, Dementin S, Burlat B, Cournac L, Volbeda A, Champ S, Martin L, Guigliarelli B, Bertrand P, Fontecilla-Camps J, Rousset M, Léger C (2008) Proc Natl Acad Sci U S A 105:11188–11193
84. Liebgott P-P et al (2010) Nat Chem Biol 6:63–70
85. Greco C, Fourmond V, Baffert C, Wang P-H, Dementin S, Bertrand P, Bruschi M, Blumberger J, de Gioia L, Leger C (2014) Energy Environ Sci 7:3543–3573
86. Millo D, Hildebrandt P, Pandelia M-E, Lubitz W, Zebger I (2011) Angew Chem Int Ed 50:2632–2634

87. Ly HKK, Sezer M, Wisitruangsakul N, Feng J-JJ, Kranich A, Millo D, Weidinger IM, Zebger I, Murgida DH, Hildebrandt P (2011) FEBS J 278:1382–1390
88. Fourmond V, Burlat B, Dementin S, Sabaty M, Arnoux P, Etienne E, Guigliarelli B, Bertrand P, Pignol D, Léger C (2010) Biochemistry 49:2424–2432
89. Ackrell BAC, Armstrong FA, Cochran B, Sucheta A, Yu T (1993) FEBS Lett 326:92–94
90. Abou Hamdan A, Dementin S, Liebgott P-P, Gutierrez-Sanz O, Richaud P, De Lacey AL, Roussett M, Bertrand P, Cournac L, Léger C (2012) J Am Chem Soc 134:8368–8371
91. Sezer M, Millo D, Weidinger IM, Zebger I, Hildebrandt P (2012) IUBMB Life 64:455–464
92. Léger C, Elliott SJ, Hoke KR, Jeuken LJC, Jones AK, Armstrong FA (2003) Biochemistry 42:8653–8662
93. Hirst J (2006) Biochim Biophys Acta 1757:225–239
94. Vincent KA, Parkin A, Armstrong FA (2007) Chem Rev 107:4366–4413
95. Jeuken LJ (2009) Nat Prod Rep 26:1234–1240
96. Lojou E (2011) Electrochim Acta 56:10385–10397
97. Butt JN (2014) Biochem Soc Trans 42:47–51
98. Ash PA, Vincent KA (2012) Chem commun 48:1400–1409
99. Compton RG, Banks CE (2011) Understanding voltammetry. Imperial College Press, London

Adv Biochem Eng Biotechnol (2017) 158: 43–74
DOI: 10.1007/10_2015_5011

Published online: 10 August 2016

Structure and Modification of Electrode Materials for Protein Electrochemistry

Lars J.C. Jeuken

Abstract The interactions between proteins and electrode surfaces are of fundamental importance in bioelectrochemistry, including photobioelectrochemistry. In order to optimise the interaction between electrode and redox protein, either the electrode or the protein can be engineered, with the former being the most adopted approach. This tutorial review provides a basic description of the most commonly used electrode materials in bioelectrochemistry and discusses approaches to modify these surfaces. Carbon, gold and transparent electrodes (e.g. indium tin oxide) are covered, while approaches to form meso- and macroporous structured electrodes are also described. Electrode modifications include the chemical modification with (self-assembled) monolayers and the use of conducting polymers in which the protein is imbedded. The proteins themselves can either be in solution, electrostatically adsorbed on the surface or covalently bound to the electrode. Drawbacks and benefits of each material and its modifications are discussed. Where examples exist of applications in photobioelectrochemistry, these are highlighted.

Keywords Bioelectrochemistry, Carbon, Electrode surface, Electrodes, Gold, Graphite, Indium-tin oxide (ITO), Mesoporous electrodes, Protein electrochemistry, Self-assembled monolayers, Surface modification

Contents

L.J.C. Jeuken (✉)
University of Leeds, Leeds, UK
e-mail: l.j.c.jeuken@leeds.ac.uk

1 Introduction

This chapter is concerned with electrode materials for bioelectrochemical and photobioelectrochemical applications. The choice of electrode material is paramount for any electrochemical experiment and this also applies to photobioelectrochemistry. The choice of electrode material is based on many factors, including price, robustness, manufacturability, reproducibility, stability and, most important of all, its surface chemistry.

Bioelectrochemistry has different meanings in scientific literature and can encompass cell membrane potentials, electrochemistry of biologically relevant metabolites (e.g. dopamine or glucose) and electrochemistry of redox proteins. The latter, which is also known as protein electrochemistry, is the primary focus of this book and this chapter. Protein electrochemistry can be further classified depending on two parameters.

First, electron transfer between electrode and redox protein can be direct or mediated. Electron transfer can be mediated by native enzyme substrates (e.g. NADH or H_2O_2) or by non-biological redox compounds (e.g. ferricyanide or methylviologen). The choice of electrode material and its modifications influences both types of systems. However, direct electron transfer is often more challenging and requires a tight control of the surface chemistry. Therefore, although this chapter presents some examples of mediated electron transfer, the focus lies on direct electron transfer.

Second, protein electrochemistry can be performed on redox proteins which diffuse freely in solution or are adsorbed irreversibly on the electrode surface. It is important to note here that, in the early years of protein electrochemistry, adsorption of redox proteins was seen as a problem that needed to be overcome for electrochemistry of freely diffusing proteins to be successful. Indeed, the first electrochemical experiments on cytochrome *c* in the 1970s were focussed on reducing the adsorption of proteins on surfaces [1]. Later, however, it was realised that electrochemistry could also be performed on proteins that are irreversibly adsorbed onto the electrode surface as long as the native fold of the protein is not perturbed upon adsorption.

Electrochemistry of adsorbed redox proteins that exchange electrons directly with the electrode is also known as protein-film electrochemistry (PFE) or protein-film voltammetry (PFV). In PFE, the orientation of the redox protein is an important factor as redox active co-factors of proteins are generally buried inside the protein and typically only approach the surface of the protein at a specific location. The electron transfer rate decreases exponentially with distance and, thus, rapid electron transfer requires the redox active co-factors to face the electrode surface. The surface chemistry of the electrode affects the orientation of the adsorbed protein and thus the electrochemical response of many proteins.

This chapter compares two commonly used electrode materials in protein electrochemistry (carbon-based and gold electrodes) and describes their various surface modifications. Furthermore, as photoelectrochemistry may require transparent electrodes, (semi)conductive transparent materials such as indium-tin oxide are discussed. Where examples exist of biophotoelectrochemical applications, mainly with photosystem I (PSI) or photosystem II (PSII), these examples are briefly covered with respect to the properties of the electrode material. However, for a more in-depth coverage of photobioelectrochemistry, we refer the reader to Chaps. 4–6 of this book. This chapter is not intended to give a comprehensive review of electrode materials, but to provide a tutorial overview, noting drawbacks and benefits of different materials and methods. We hope this chapter gives readers a basis on which to select their preferred electrode material.

2 Carbon Based Materials

By far most of the electrode materials used in bioelectrochemistry are carbon based. Most carbon materials, with the exception of diamond, are cheap and abundant and can be easily produced in a variety of different structures, from hard, solid materials to fine powders which can be screen printed. Many carbon materials have been used for protein bioelectrochemistry, in particular materials based on the graphene carbon structure, such as highly-orientated pyrolytic graphite (HOPG), pyrolytic graphite, glassy carbon, carbon nanotubes, fullerenes and graphene (nanoflakes). Other non-graphite materials include doped diamond and carbon paste.

2.1 Basal Plane Electrodes and Highly-Orientated Pyrolytic Graphite (HOPG)

In pyrolytic graphite, the planar graphene sheets are oriented to give a material with a well defined basal and edge plane (Fig. 1). Because of the anisotropic nature of the material, the electrical resistivity is different when measured either parallel or perpendicular to the graphene plane. For basal plane electrodes the side of the graphene layer is used as the electrode's surface. Basal plane electrodes are prepared by stripping or cleaving one or several layers of graphene, exposing a

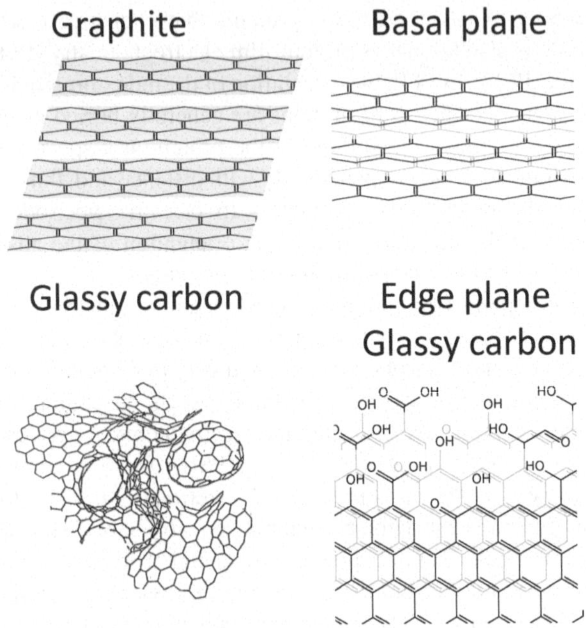

Fig. 1 *Left:* Structure of two common carbon electrodes, graphite (pyrolytic graphite or highly-ordered pyrolytic graphite, HOPG) and glassy carbon. The exact structure of glassy carbon is not exactly known. The structure shown here is based on a proposed structure by Harris et al. [15]. *Right:* Schematic representation of the surface chemistry of basal plane and edge plane pyrolytic graphite electrodes. The basal plane is hydrophobic and exposes the side of the sp^2 polycyclic aromatic graphene layers from graphite. In contrast, the edge of the polycyclic aromatic graphene layers is exposed in pyrolytic graphite edge (PGE) plane electrodes. The breaking of the polycyclic aromatic structure results in a variety of oxygen-carbon based organic groups that give the surface hydrophilic properties. The surface of glassy carbon exposes a mixture of basal and edge planes

fresh and clean surface in the process. Freshly cleaved basal plane electrodes can simply be prepared by gently pressing down single-sided Scotch tape and gently pulling off the tape, striping the top layers of graphene in the process. In highly-orientated pyrolytic graphite (HOPG), the anisotropic nature of the graphite is maintained to an almost perfect degree, providing an almost defect-free basal plane surface. Although HOPG has the advantage that the basal plane surface is atomically flat, basal plane electrodes are hydrophobic and generally not suitable as a bioelectrochemical interface. Nonetheless, studies have been performed with HOPG and basal plane electrodes, mainly because the atomically flat surface of the basal plane can have important experimental advantages. For instance, two studies reported by the same laboratory studied horseradish peroxidase [2] and catalase [3] on HOPG with electrochemical scanning tunnelling microscopy (ECSTM). It was found that the hydrophobic basal plane resulted in poor adsorption behaviour of catalase and clusters of aggregates were observed [3]. A similar result was obtained using atomic force microscopy (AFM), which showed that laccase forms large aggregates on HOPG, leaving large parts of the surface devoid of protein [4]. In spite of these problems, PSII was shown to form 2D crystals on HOPG, although it was noted that the reproducibility was limited by non-uniformity and amorphisation [5]. Where highly-ordered crystals were formed, consistent STM images were obtained which indicated spatially-coherent electron tunnelling through single PSII proteins.

For bioelectrochemical purposes, the basal plane surface requires chemical modification, either covalently or non-covalently. An example of covalent modification is controlled anodisation under mildly electrochemical oxidation conditions, which forms a thin oxide layer which is a more suitable surface for the adsorption of protein [3]. To maintain the smooth geometric surface of the basal plane, these modifications need to be accurately controlled as further anodisation leads to rough surfaces, which are arguably not that different from edge plane electrodes (see Sect. 2.2).

2.1.1 Non-covalent Modification of Basal Plane Pyrolytic Graphite

In the 1990s, Rusling explored the use of films of amphiphiles (detergents or lipid analogues) forming multilayer films on basal plane electrodes (Fig. 2a) [6]. Films were cast onto the solid surfaces from aqueous vesicle dispersions mixed with protein, followed by the evaporation of the solvent. Heme proteins such as myoglobin readily displayed electro-active signals, suggesting the proteins diffuse between the multilayers on the electrode surface. In particular, the surfactant didodecyldimethylammonium bromide (DDAB) was explored for these cast films, but other surfactants also produced well-behaved reversible signals. An alternative method to casting a film is a layer-by-layer (LbL) deposition of protein and polyions such as polystyrenesulfonate (PSS), where the basal plane electrode is alternatively incubated in a PSS and a protein solution. Later, concerns were raised that these amphiphilic layers released the heme of a small part of the proteins and the bioelectrochemical signals were in fact produced by free heme [7, 8]. The initial experiments by Rusling used spectroscopy to confirm that the heme proteins retain their native structure when in the DDAB film. However, if intact heme proteins are electrochemically 'silent' in this system, it is possible that the electrochemical signal is instead obtained from a small amount of free heme which is not detected by spectroscopy. The issue is still being debated and the nature of the electrochemical signals might well depend on environmental parameters, such as pH (for a discussion, see [9]). Nonetheless, the use of these cast films illustrates an important point: if your immobilised enzyme is electrochemically inactive, even small amounts of impurities or denatured protein could be responsible for observed electrochemical signals.

2.2 Pyrolytic Graphite 'Edge' (PGE) and Glassy Carbon (GC) Electrodes

Basal plane electrodes contain so-called defects, where the graphene top layer is broken and some of the 'edges' of the graphene layers are exposed to the surface.

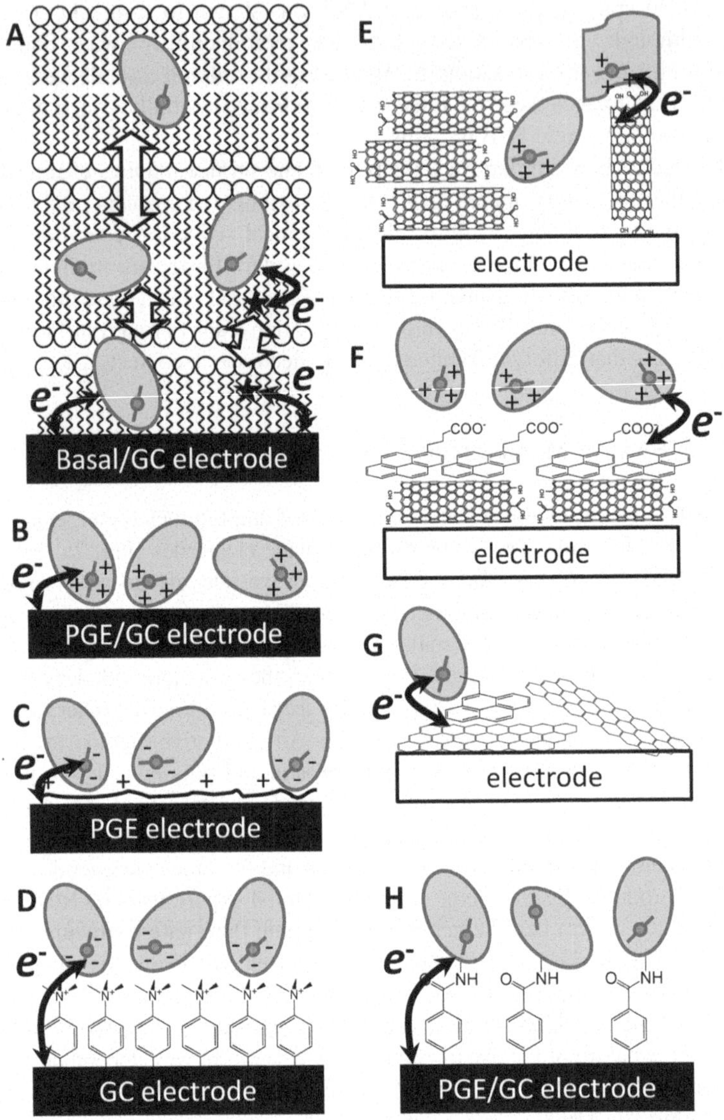

Fig. 2 Schematic representations of different methods of preparing immobilised protein layers on carbon electrodes. The *solid arrows* represent electron transport and the *open arrow* in (**a**) represents diffusion of either proteins or mediators. (**a**) Proteins are incorporated in amphiphilic multilayers. The protein either diffuses between layers to and from the electrode or mediators are added to the amphiphile layers to transport electrons between proteins and the electrode surface. These surfaces have been prepared on basal plane graphite and glassy carbon (GC) electrodes. (**b**) Direct adsorption of proteins on pyrolytic graphite edge (PGE) and GC electrodes. Proteins are physisorbed on the electrode via a combination of hydrophobic and electrostatic forces. (**c**) The use of promotors to aid the physisorption to PGE electrodes. Cationic promotors can use the charge of the PGE electrode to aid in the adsorption of negatively charged proteins or change the orientation of proteins with positive and negative 'patches' on their surface. (**d**) Covalent modification of GC electrodes is used to tailor the surface chemistry for protein adsorption. In this example, a trimethylethanammonium group is attached to the GC surface to provide a positive

Direct comparison between freshly cleaved basal plane electrodes and polished pyrolytic 'edge' electrodes led to the conclusion that redox proteins preferentially interact with these oxygen-containing defects [10]. Thus, care needs to be taken when evaluating bioelectrochemical studies that use basal plane pyrolytic electrodes, as it is possible that interfacial electron transfer is likely to occurs at the defects. A direct comparison with so-called pyrolytic graphite 'edge' (PGE) or glassy carbon (GC) electrodes can be performed to provide insight into the electrochemical response.

In contrast to basal plane electrodes, PGE proved to be a very suitable electrode material for bioelectrochemistry and is now used routinely. PGE electrodes can be easily pre-treated by polishing or abrasion to remove previously adsorbed material and impurities, providing a fresh surface with many oxygen-containing functionalised groups (Fig. 1) on which many redox proteins readily adsorb in a native state. Unlike the basal plane, the edge plane is microscopically very rough, leading to an increase in electrochemical surface area [11]. The electrochemical surface area is dependent on the polishing method and polishing with microparticles (e.g. 1-μm alumina) leads to a larger surface area than when nanoparticles are used (e.g. 50-nm alumina). However, polishing with microparticles does not automatically lead to increased protein loading as the rougher surfaces expose relatively more of the basal plane of graphene layers [11]. The material used to polish the electrode (e.g. alumina and diamond polish) has been found to remain on the electrode surface even after rinsing electrodes or incubating them in an aqueous solution in a bath sonicator. Therefore, it cannot be excluded that the polishing material contributes to the observed bioelectrochemical behaviour of the electrode material.

The negative charges of the oxygen-containing functional groups of PGE surfaces are very suitable for the electrochemistry of positively charged proteins (Fig. 2b). For instance, cytochrome *c* shows good electrochemical behaviour when adsorbed on PGE [12]. It is well known that cytochrome *c* contains a number

Fig. 2 (continued) charge to aid the binding of negatively charged proteins. (**e**) Coating of electrodes with carbon nanotubes. Single-walled nanotubes are shown, but multi-walled nanotubes are also used. The hydrophobic sides of the nanotubes are hydrophobic and thought not to interact favourably with proteins. This schematic representation shows the proteins interacts with the ends of the tubes, although activating the sides of multi-walled carbon nanotubes is also possible. Because of the small diameter of carbon nanotubes, it is possible they are able to approach occluded pockets of proteins that are normally not accessibly to macroscopic electrodes (**f**) Modification of hydrophobic sides of carbon nanotubes (single- or multi-walled) with polycyclic aromatics is shown. This modification creates surface chemistries that exhibit more favourable interactions with proteins. (**g**) Modification of electrode surface with graphene 'flakes'. As with the carbon nanotubes, the side of the carbon nanoflakes is hydrophobic and less suitable to interact with proteins. This scheme represents a method in which the protein is covalently modified with a polycyclic aromatic compound that binds to the side of the graphene flakes via π–π interactions. (**h**) Proteins can be covalently bound to PGE or GC electrodes. In this example, the carbon electrode is modified to contain carboxylic acid groups, which are then coupled to protein amines via the formation of peptide bonds (using, for instance, EDC/NHS (1-ethyl-3-(3-dimethylaminopropyl)-carbodiimide/*N*-hydroxysulfosuccinimide))

of positively charged lysines around its heme edge, which results in a favourable interaction with negatively charges electrodes. This is confirmed by their interaction with gold electrodes that have been modified with carboxylic acid terminated monolayers (see Sect. 3.1) [13]. It is not necessary for a protein to have a net positive charge to adsorb onto PGE, as electrostatics is not the only force to control adsorption – hydrophobic interactions also contribute.

Because PGE also exposes a significant amount of basal plane surface [11], it can be expected that proteins preferring a hydrophobic surface also interact favourably with PGE. As already noted above for HOPG, this is relatively rare and most proteins tend to denature on very hydrophobic surfaces. A well studied exception is the blue copper protein azurin, which shows good electrochemical behaviour on gold electrodes modified with hydrophobic monolayers and, indeed, on PGE electrodes [14].

Besides PGE electrodes, glassy carbon (GC, also known as vitreous carbon) electrodes are widely used in bioelectrochemistry. GC is extremely inert and resistant to high temperatures and chemical treatments. Similar to graphite, GC is made up of sp^2 bonded carbon (without sp^3), but the structure is very disordered (Fig. 1) [15]. The surface of GC is thus less defined, but the surface chemistry of GC electrodes can be thought of as a mixture of basal and edge planes. As with PGE, the GC electrodes are typically cleaned by polishing with alumina or diamond polish with which a mirror-like shine can be obtained. To increase the hydrophilicity of the GC surfaces, they can be anodised with acid [16] or activated by exposing the surface to a methane flame [17]. Activated GC behaves in many respects similar to PGE, although small differences are observed [18].

2.2.1 Non-covalent Modification of PGE and GC

Many redox proteins do not give an electrochemical response on PGE because they denature, do not adsorb on negative surfaces or adopt an orientation unfavourable for interfacial electron transfer. In these cases, the PGE surface can be modified, either covalently or non-covalently. One of the simplest non-covalent modifications, initially exploited by Armstrong and co-workers for PGE, is by incubation with metal cations [10], organic cations [19, 20] or polycations [21] (Fig. 2c). (Poly) cations readily adsorb to PGE, in effect changing the negative surface charge to positive. The (poly)cations for electrode modification are also known as promotors as they promote the electron transfer between the electrode and the redox proteins. Examples of often used and commercially-available (poly)cations are polylysine, neomycin and polymyxin, but many other (poly)cations such as chitosan are also commonly used [18]. The PGE electrodes can either be immersed in a (poly)cation solution before rinsing and further incubation with the protein or the (poly)cation can be mixed with the target protein before incubating the electrode. Because of the rough surface of PGE, it not exactly known what the structure of the polycation-modified surface looks like. Some depictions of the polycation/protein surface visualise the polycation as a layer many nanometres thick in which the protein resides. However, this is unlikely as, from the few cases where the polycation/protein layers have been studied in more detail (e.g. Forzani et al studied the layer-by-layer assembly of polycations with polyphenol oxidase [22]), it is clear that the polycation layer is a relatively thin layer (compared to the protein size) onto which a protein layer adsorbs (Fig. 2c). The cartoon representation in Fig. 2c also corresponds to the use of small metal or organic cations, which only form a thin or even (sub)monolayer film on the surface. In many cases, direct electron transfer from the PGE electrode to the redox protein in the polycation/protein layer is possible. This indicates that the polycation does not perturb the coupling between the redox protein and the surface and thus either the protein intercalates into the polymer and is simultaneously in direct contact with the electrode surface or the polycation only forms a layer of several Ångstrom between the electrode and the protein.

Other modifications, such as with bentonite clays, have also been explored. For instance, by applying clay mixtures to roughly polished basal plane electrodes (exposing the edge planes), reversible electrochemical signals were obtained for cytochrome *c* [23]. Clays are aluminium/silicon oxide sheets that are negative charged and provide a suitable microenvironment for the positively charged cytochrome *c*. In general, it has been observed that clays have beneficial properties as a promotor for heme proteins such as cytochrome c, cytochrome P450 and myoglobin.

Similar to basal plane graphite electrodes, non-activated GC (i.e. not treated by acid or heat) has been used to cast multilayer amphiphile films (Fig. 2a). The group of Rusling recently prepared 2-μm multilayer lipid films and incorporated PSII into these layers [24]. High photocurrents were only obtained when an electron mediator was added, which was attributed to the fact that PSII might not transfer electrons efficiently to the underlying GC electrode. However, it is likely that diffusion of PSII through the lipid bilayers is slow, if not impossible, which might be an alternative explanation of why electron mediators are required in this setup.

2.3 Covalent Modification of Graphite Electrodes

Covalent modification of graphite electrodes potentially has the ability to provide a wider variety of surface chemistry than non-covalent modification. Ideally, covalent modification of graphite should provide a single monolayer of a chemical compound that contains a 'head' group pointing away from the surface into the electrode. The head group should exhibit specific chemical properties for further interaction and/or adsorption of the redox proteins.

Graphite is relatively unreactive and covalent modification thus relies on radical chemistry. The most common covalent modification is performed with aromatic diazonium salts, e.g. benzenediazonium cation (Fig. 2d). The electrochemical reduction of aryl diazonium salts produces radical anions, which react with the carbon surface to form covalent bonds. As described in the tutorial review by

Pinson and Podvorica [25], the amount of radicals formed electrochemically can be quantified by integration of the reduction current, although this analysis does not provide insight into what fraction of the radicals react with the carbon electrode to form a covalent bond. Electrochemical analysis of experimental data suggests that the number of radicals formed by the reduction of aromatic diazonium salts could form a close-packed monolayer on the surface. However, detailed analysis with atomic force microscopy (AFM) and other surface analytical techniques suggests instead that only submonolayers are formed which are, in places, several layers thick. It is thought that multilayers are formed because radical anions react with aryl molecules already attached to the surface [25, 26].

The formation of multilayers can be prevented by placing bulky groups on the *meta*- and/or *para*-positions of the benzenediazonium cations [27, 28], but such modification severely limit the type of surface chemistry with which the surface can be modified or requires the use of more complex chemistry to remove the bulky groups afterwards. Because of the chemical complexity associated with the synthesis of 'protected' diazonium salt and subsequent removal of these protecting groups, its application to bioelectrochemistry has been limited, although some interesting examples have been reported where (sub)monolayers of redox active groups such as ferrocene and quinones were grafted on carbon electrodes [29, 30].

The modification with diazonium salts has also been exploited in microbial fuel cells, where the surface chemistry of the carbon was optimised for the electrochemical communication between the electrode (anode) and bacteria in a biofilm grown on the carbon surface [31, 32]. Finally, diazonium salts have been used to modify surfaces with substrate analogues, which then bind to enzymes, contributing to their binding to the surface, their orientation on the surface and providing an efficient electron-transfer pathway to the enzyme's active site. An example of the latter approach is the modification of graphite surfaces with anthracene compounds which bind efficiently to laccases [33, 34].

2.4 Carbon Nanotubes and Graphene

As the research into carbon nanotubes increased rapidly in the 1990s, their potential benefits in protein electrochemistry were explored [35]. Carbon nanotubes, in principle, can enhance the loading of a redox protein on an electrode surface by greatly enhancing the electroactive area of the carbon electrode. Furthermore, these nanomaterials have the ability to interact closely with occluded pockets of a protein surface, which might otherwise be inaccessible to macroscopic flat electrodes (Fig. 2e). In this respect it is important to note that pyrolytic graphite 'edge' electrode also have a very rough surface with many nanoscale features and thus could share some properties with electrodes modified with carbon nanotubes or nanoflakes (nanosized fragments of graphene).

The side walls of carbon nanotubes (either single- or multi-walled) are hydrophobic and, similar to basal plane electrodes, not typically suitable for bioelectrochemical application. In fact, redox enzymes adsorbed on the side wall carbon nanotubes are likely to denature [36]. Similar to basal plane electrodes, approaches have been reported to solve this problem, such as the use of surfactant films [37], or the wall can be activated electrochemically [38] with plasma [39] or by acid treatment [40]. When heme proteins are used, the electrochemical response in the presence of surfactants needs to be critically assessed to ensure that the surfactant has not released the heme groups from the proteins [7, 8]. If single-wall carbon nanotubes are used, plasma, electrochemical or acid treatment opens up the tubes and their surface is oxidized to create carboxylic, carbonyl and hydroxyl groups, producing a very similar surface chemistry to that of glassy carbon or edge plane pyrolytic graphite electrodes. Alternatively, non-covalent modification with pyrenyl compound (or other polycyclic aromatic compounds) is a relatively simple and promising approach (Fig. 2f) [41–43]. π–π interactions between polycyclic compounds and the side walls of the nanotubes are strong and, when, for instance, hydroxy or carboxylic-acid derivatives of polycyclic aromatic compounds are used, a suitable surface chemistry can be introduced to the nanotubes. In principle, the same modification is possible for basal plane and HOPG electrodes (Sect. 2.1.1), but few reports have appeared on this topic. A rare example of the latter is an approach in which surface-exposed cysteines of a cytochrome P450 are chemically coupled to a pyrene, which is then used to immobilise the protein to basal plane electrodes [44]. The advantage of this method is that the site-specific modification of the redox proteins enables control over the orientation on the surface. Finally, carbon nanotubes have been covalently modified with diazonium salts to create and control positively and negatively charged surfaces and thus the interaction with enzymes [45].

Carbon nanotubes have also been used in an attempt to enhance the electronic coupling with whole organisms in microbial electrochemistry. A photoelectrochemical example is the use of nanotubes to enhance the current output of immobilised cyanobacteria [46].

Similar to carbon nanotubes, most of the surface of graphene 'nanoflakes' is akin to that of basal plane electrodes and typically unsuitable for bioelectrochemistry unless activated to graphene oxide, which is defined as submicron sheets of graphene with many carboxylic, carbonyl and hydroxyl moieties. Heme proteins mixed with graphene oxide can be deposited on electrodes to give rise to reversible electrochemical signals [47]. Although the proteins retain their structural integrity, it is yet to be shown that graphene oxide modification has beneficial effects over the use of PGE and GC electrodes, other than increasing the effective surface area of the electrode. Finally, in rare cases, reports have emerged of the use of 'activated' fullerenes for bioelectrochemical purposes [48]. As fullerenes are themselves redox active, their action might be more like that of a mediator.

When deposited on transparent materials, continuous thin graphene films have an important potential benefit for photobioelectrochemistry in that they are transparent. Some work has recently been reported in which graphene electrodes are employed in photobioelectrochemical applications. Feifel et al. deposited PSI on graphene electrodes either by modifying the graphene with polycyclic aromatic compounds or by modifying PSI with a pyrene group (Fig. 2g), i.e. similar approaches to those above for carbon nanotubes [49]. In this particular example, graphene was deposited on polished silicon wafers, which are not transparent, but there is no reason why the same modification cannot be carried out on transparent materials.

2.5 *Covalent Attachment of Proteins to Graphite Surfaces*

It has already been noted that in the early years of protein bioelectrochemistry it was thought that adsorption of proteins on electrode surfaces was considered a problem as this impaired electron transfer between proteins freely diffusing in solution and the electrodes. Later it was recognised that electrochemistry of proteins irreversibly adsorbed on the electrode was a very powerful method, not limited or convoluted by diffusion kinetics of the protein and requiring less protein sample. In the previous sections, many examples are provided of how proteins can be adsorbed to surfaces and, by optimising conditions (e.g., temperature and the use of promoters), stable protein films have been created for the duration of the experiment. However, in other examples, the electrochemical signal deteriorates during the course of the experiment, which could be because of loss of the protein film by denaturation of the protein or its desorption. In the case of the latter, covalent immobilisation of the protein to the surface can offer a solution. Covalent attachment to electrodes is possible by forming a layer of amine groups on the surface with the use of aromatic diazonium salts (see Sect. 2.3), followed by chemical coupling between the surface-amine and carboxylic acids groups on the surface of the protein, forming a peptide bond. This chemical coupling can, for instance, be performed by common coupling reagents (e.g. EDC/NHS; 1-ethyl-3-(3-dimethylaminopropyl)-carbodiimide/*N*-hydroxysulfosuccinimide). The 'opposite' coupling, where the surface is modified with carboxylic acids, which are then coupled to amines on the proteins, is also possible. Covalent attachment has been reported for a number of enzymes, such as hydrogenases [50, 51], glucose oxidase [52] and peroxidase [53]. A drawback of coupling the enzymes to the surface via EDC/NHS is that the orientation of the enzymes is not controlled, similar to the typical non-covalent immobilisation of enzyme described in the previous sections. This might limit the fraction of the enzymes that efficiently exchange electrons with the electrode.

3 Gold Electrodes

After carbon, gold is probably the most commonly used electrode material for bioelectrochemistry. 'Bare' gold is rarely used as electrode material and, instead, gold is popular because it can relatively easily be modified with sulphur-containing compounds using self-assembly. The typical example is the modification of gold with long alkane-thiols, where the thiol group binds irreversibly to the gold surface, creating a so-called self-assembled monolayer (SAM) of alkane-thiols.

3.1 Covalent Modification of Gold Electrodes: Self-Assembled Monolayers

When clean gold electrodes are incubated (typically for 16–24 h) in a solution containing alkane-thiols (typically 1 mM), the alkane 'tails' form a densely packed monolayer (Fig. 3). Because the cross-section of the carbon chain is less than the distance between sulphur atoms on a gold lattice, the alkane chains adopt an angle with respect to the gold surface (28° for Au[111] surface on which thiols adopt a ($\sqrt{3} \times \sqrt{3}$)R30° ordering [54]). The lateral van der Waals interactions between the carbon chains help to stabilise the SAM, which, for long alkanes such as octadecane-thiol, can be considered a permanent modification of the gold, in spite of the somewhat labile nature of the coordinate bond between the sulphur and the gold. SAMs can also be made from almost any thiol or dithiol compounds, including smaller alkanes (e.g. 2-mercapto-ethanol) or phenyls (e.g. 4-mercapto-phenol), but these SAMs are generally less stable, particularly under electrochemical conditions. For an extensive review of SAMs on gold and other surfaces, we refer the reader to Love et al. [54] and references therein.

The formation of SAMs on gold is primarily used to control the surface chemistry, where, for instance, alkane-thiols form a hydrophobic surface and ω-hydroxy-alkane-thiols can form a hydrophilic surface (Fig. 3a). Similarly, positively-charged surfaces are typically made with ω-amino-alkane-thiols and negative charges can be introduced with ω-mercapto-alkanoic acid (Fig. 3a). The important benefit over the chemical modification of carbon (see Sect. 2.3) is that the formation of SAMs on gold is relatively easy to control, without the danger of forming submonolayers or multilayers. It is believed that the somewhat labile gold–sulphur bond allows for migration of a thiol across the surface, which in turn enables the ideal packing of SAMs.

As the surface chemistry influences the adsorption of proteins and the orientation with which redox proteins interact with the electrode, either within an immobilised protein film or transient interaction for diffusion electrochemistry, different SAMs show different behaviour in protein electrochemistry (Fig. 4a, b). A classical example is cytochrome *c*, which adsorbs on negatively charged surfaces [55]. By mixing carboxylic acid terminated thiols (negatively charged) with alkanethiols, the total charge on the surface can be tuned. The electron transfer kinetics of cytochrome *c* improves with mixed SAMs, likely because the charge density influences the protein orientation and/or mobility within the protein film [56]. Hydrophobic surfaces prepared from alkanethiols alone are only used in rare cases [57, 58]. In our laboratory we generally use combinations of methyl-, hydroxy-, amino- and carboxylic-acid-terminated thiols to compare different

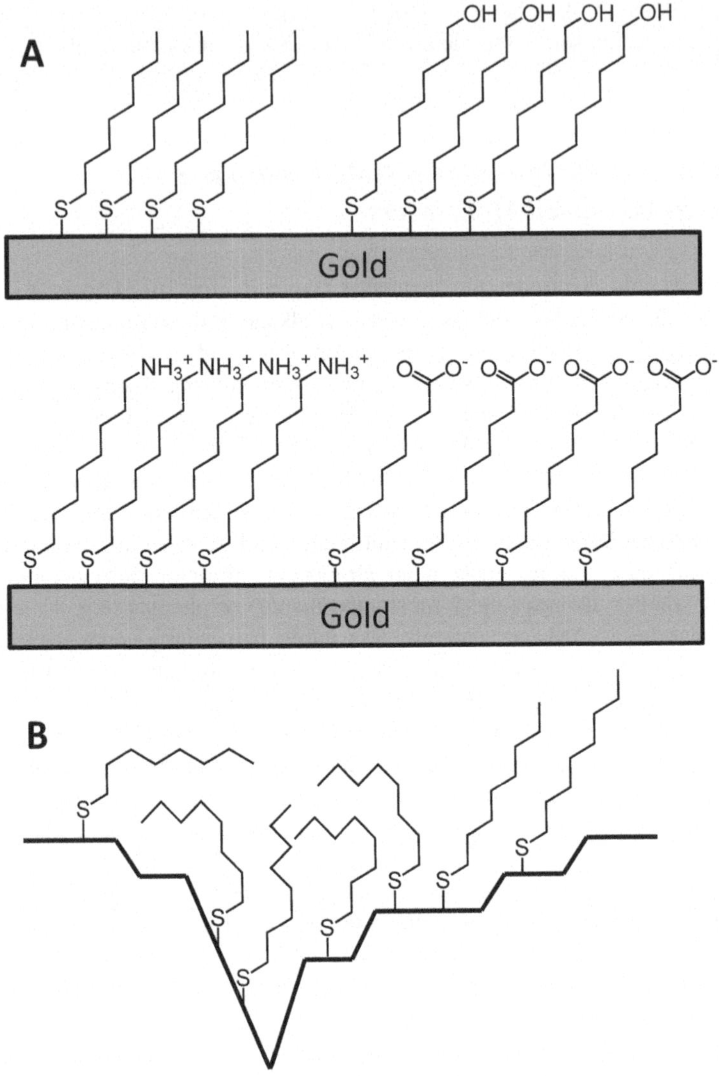

Fig. 3 Schematic representations of gold electrodes modified with a self-assembled monolayer (SAM) of alkanethiols. (**a**) Four examples of SAMs on smooth gold surfaces (e.g. Au[111]), where the tilt-angle of the alkane chains is known to have a angle of 28° with respect to the normal of the gold surface. Four different alkanethiols are shown which provide a hydrophobic (1-octanethiol), a hydrophilic (8-hydroxy-1-octanethiol), a positively charged (8-amino-1-octanethiol) or a negatively charged (8-mercapto-1-octanoic acid) surface. For applications in protein bioelectrochemistry a mixture of thiols is usually required for optimal electron transfer. (**b**) Schematic representation of a disordered SAM of 1-octanethiol on a rough gold surface

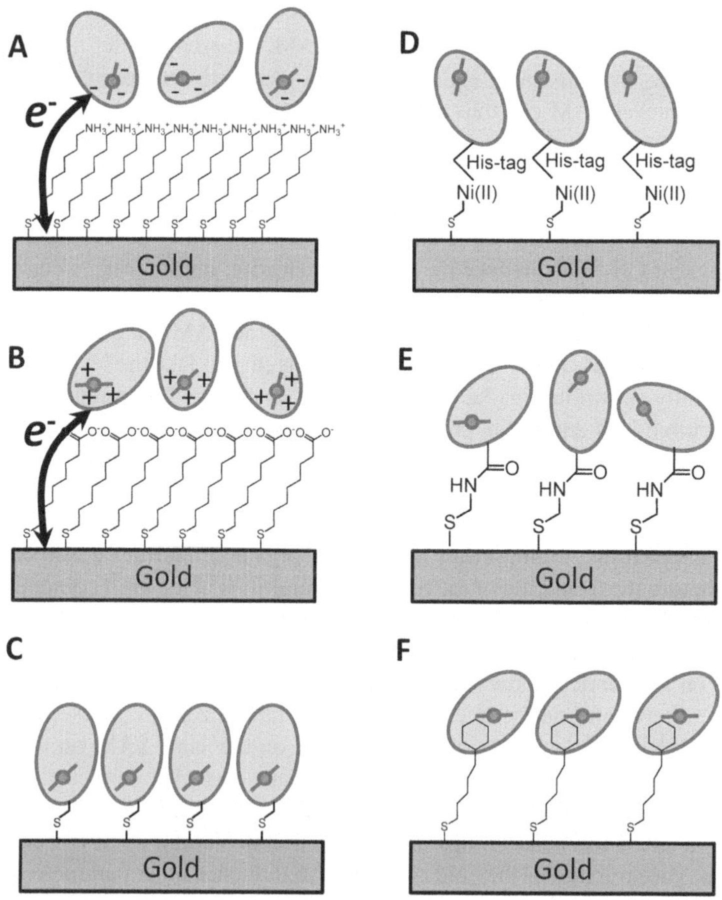

Fig. 4 Schematic representations of different methods of preparing immobilised protein layers on gold electrodes. (**a**) SAM of 8-amino-1-octanethiol can be used to bind a negatively-charged (patch of a) protein to the gold surface. (**b**) SAM of 8-mercapto-1-octanoic acid can be used to bind a positively-charged (patch of a) protein to the gold surface. (**c**) Cysteine residue on the surface of a protein can be used to directly couple the protein to bare gold where it forms an Au–S bond. (**d**) Engineered polyhistidine-tags, commonly used to purify proteins, can be used to bind the protein to a gold surface modified with a SAM containing, for instance, nitrilotriacetic acid (NTA) groups binding a Ni(II) or Co(II). (**e**) Proteins can be chemically coupled to SAMs on gold surfaces. In this example, a SAM is formed containing amine groups that are chemically coupled to carboxylic acids groups on the surface of a proteins by forming a peptide bond (using, for instance, EDC/NHS (1-ethyl-3-(3-dimethylaminopropyl)-carbodiimide/*N*-hydroxysulfosuccinimide) as chemical reagents). (**f**) Proteins have been 'wired to' or 'plugged into' gold surfaces by modifying the surface with either a co-factor or substrate homologue that binds tightly to the proteins

surface chemistries for bioelectrochemistry (for example, see [59, 60]). Often, mixtures of thiols, rather than SAMs made of single thiols, show the fastest interfacial electron transfer kinetics [56, 59, 60].

By changing the chain lengths of alkane-thiols, a second feature which can be systematically altered is the 'thickness' of the SAM. Electron transfer rates exponentially decrease with distance, e.g. between the electrode and redox centre in the protein. A thicker SAM can thus drastically reduce the interfacial electron transfer rate. At very short distances, other processes limit the electron transfer rate and decreasing the SAM thickness below a certain limit does not improve the electron transfer kinetics. Because SAMs from short thiols (e.g. 2-mercaptoethanol or 4-mercaptobutanol) are less stable and SAMs from long thiols (above chain lengths of about eight CH_2 units) reduce the interfacial electron transfer rate, a compromise has to be made. Typically, we make SAMs from hexane- or octane-thiols.

A second feature that influences the thickness of the SAM is the roughness of the gold electrode. SAMs prepared on very smooth electrodes take the ideal structure as schematically shown in Fig. 3a. In contrast, SAMs prepared on rough surfaces are less structured [61] and redox proteins can approach the gold surface further than predicted based on the length of the alkanethiol (Fig. 3b). Non-structured areas in a SAM are often denoted as defects, although for rough surfaces the defect density can be high and a defining feature of the SAM. Besides surface roughness, defects in SAMs appear due to impurities in the gold, organic impurities adsorbed on the surface before the formation of the SAM and impurities in the thiol compound used to form the SAM. The structure of the SAM affects the interfacial electron-transfer rate. For instance, the electron transfer rate between azurin (a mono-copper redox protein) on an atomically flat Au[111] surface modified with decanethiol has been reported to have an interfacial electron-transfer rate at zero over-potential (k_0) of $290 \pm 20\ s^{-1}$ [58]. In contrast, azurin adsorbed on the same SAM on rough gold electrodes shows a faster electron transfer rate between 400 and 470 s^{-1} [14, 57]. Disorder in SAMs also influences the surface chemistry because in a disordered SAM the head-groups are not structurally arranged at the electrolyte interface. The structure of the gold can thus also influence other parameters such as protein adsorption (or loading of protein on the surface) and protein orientation.

One of the major drawbacks of using gold electrodes modified with SAMs is the limited potential window in which they are stable. The gold–thiol bond can both be reduced (detaching the thiol from the surface) and oxidised (presumably creating sulfur-oxides, which detach from the surface). The potential at which this occurs depends on the stability of the SAM, which – as mentioned above – depends in part on the lateral van der Waals interaction between the thiols in the SAM. Nonetheless, the electrochemical window in which most SAMs are generally stable lies between −0.4 and +0.6 V vs standard hydrogen electrode (SHE), which can be a limiting factor in many (photo)bioelectrochemical applications. Another common problem of gold electrodes are the formation of Au(I)Cl at potentials above +0.4 V vs SHE, which typically results in a sharp reduction feature at around 0.2 V vs SHE when the gold is reduced again to Au(0). If the electrolyte solution contains (traces) of chloride, this limits the potential window further to between −0.4 and +0.4 V vs SHE.

3.2 *Direct Coupling of Proteins to Gold Surfaces*

As with carbon electrodes, proteins have been directly coupled to bare gold surfaces [62–64]. Cysteines engineered on the surface of a protein (or already present in the native protein) form a strong thiol–gold bond, providing a straightforward way to adsorb and orientate a protein on the gold surface (Fig. 4c). An important potential drawback is that many proteins tend to denature on bare gold surfaces and stable proteins have to be used that can withstand the interaction with the bare metal surface. A benefit over the covalent coupling to carbon electrodes (Sect. 2.5) is that the specific engineering of cysteines on the protein surface enables control over the protein orientation and thus a way to optimise interfacial electron exchange.

An alternative method to couple proteins to (gold) surfaces is to prepare a SAM that contains specific binding groups to bind the protein. The most common method is to include a nitrilotriacetic acid (NTA)-based thiol in the mixture when forming a SAM. Together with a metal-ion, usually Ni(II) or Co(II), the NTA binds to a polyhistidine-tag (His-tag) engineered on the target proteins (Fig. 4d). His-tags are commonly used for the purification of proteins and are thus often already present in the protein of interest. There are two problems with this approach. His-tags are generally engineered on the N- or C-terminus and, therefore, this provides less control over the orientation of the protein on the surface as, for instance, the engineering of a cysteine on the surface of a protein. Furthermore, the NTA-thiol, together with the His-tag can be several nanometres in length, which is too long for efficient electron transfer. If the His-tag is rather flexible, the protein itself might retain sufficient flexibility to interact favourably with the surface, enabling electron transfer via a route that does not involve the His-tag. The Ni(II) or Co(II) used to couple the His-tag to the NTA moiety is redox active which can lead to additional redox signals, although this has also been exploited by using the metal ion to mediate electron transfer from the gold surface to the protein of interest [65].

The chemical coupling strategies already explained for carbon electrodes (Sect. 2.5) are of course also available for gold electrodes modified with amine- or carboxylic acid-containing SAMs (Fig. 4e). An example in photobioelectrochemistry is the chemical coupling of glucose oxidase to SAMs prepared from cystamine [66]. The glucose oxidase was then further chemically coupled to light-excitable ruthenium dyes. Upon illumination, electrons were proposed to transfer from the ruthenium dye to the surface, where the hole generated was filled by glucose oxidase, turning over glucose.

More elaborate schemes to couple redox proteins to a gold surface have also been described. For instance, the surface can be modified with co-factor or substrate analogues which are then coupled to an enzyme missing its co-factor (i.e. the apo-enzyme) or its substrate (Fig. 4f) [67]. This is also known as 'wiring' or 'plugging into' an enzyme. A photobioelectrochemical example of this is given by Terasaki et al. [68], who coupled PSI to a quinone modified surface.

3.3 Surface Roughness of Gold Electrodes

As explained above, the surface roughness or smoothness of gold electrodes can affect the properties of the SAM and thus the bioelectrochemical behaviour. In this section we briefly describe some commonly used gold electrodes and comment on their properties.

The simplest gold electrode is prepared from bulk gold, typically a gold rod embedded in a resin or polytetrafluoroethylene (PTFE). The gold surface is typically cleaned by polishing with alumina or diamond slurry followed by electrochemical cleaning by cycling either between −0.4 and 1.4 V vs SHE in 0.05 M H_2SO_4 or between −1.2 and −0.2 V in 0.05 M KOH [69]. The exact potential window and acid or base electrolyte composition differs somewhat between groups. The exact cleaning mechanism is unknown, but in general the cleaning procedures are believed to etch the gold surface (stripping the top layers of gold atoms) and to oxidise/reduce organic contaminants to products that adhere less well to the surface and can easily be washed off or displaced by thiols molecules during the formation of a SAM. Alternative cleaning procedures include ultrasonic cleaning, exposing the surface to UV/ozone and dipping them in chemical reagent (e.g. piranha reagent or dilute aqua regia). Often different combinations of cleaning procures are used. Because many of these procedures etch the gold surface, they influence the surface roughness, either smoothing it if the starting surface is very rough, or roughening it if the surface is very smooth (see introduction and supplementary information of Li et al. (2014) for a discussion [70]). For bulk gold electrodes, the polishing steps dominate the surface roughness and the surface roughness can be difficult to control. If the reproducibility of the gold surface becomes important, other gold surfaces need to be considered.

The creation of thin gold films with thermal evaporation or sputtering is a common method to create smooth surfaces (compared to polished bulk gold). Again, the exact conditions determine the gold surface roughness, but thermal evaporation typically results in a peak-to-peak height differences between several and tens of nanometres, although sputtered surfaces generally have a smoother surface (peak-to-peak of several nanometres). Flame annealing of these surfaces creates monocrystalline Au[111] terraces on the surface, which are atomically flat. However, the edges between the terraces are quite rough and create defects in the SAMs [54]. Finally, template-stripped gold surfaces can be very useful [71]. In this procedure, a thin layer of gold (typically 100–150 nm) is evaporated (or sputtered) on atomically flat mica or silicon wafers and a glass slide is glued to the gold layer. When the glass slides are detached from the mica or silicon wafer, the gold remains glued to the glass as gold does not adhere strongly to either mica or the SiO_2 layer of the silicon wafer. The gold surface that becomes exposed upon 'stripping' the slides retains the smoothness of the mica or silicon wafer. Although polycrystalline, this surface is very smooth with differences in peak-to-peak heights of less than 1 nm. Another benefit is that, when still attached to the mica or silicon wafer, the gold surface is protected from the air and thus remains free of organic contaminants during storage. When the glass/gold slides are stripped, the gold surface is clean and does not need further cleaning or treatment before use.

For some applications, such as surface-enhanced infra-red absorbance (SEIRA) spectroscopy, very rough gold surfaces are required, which are produced via chemical reduction of gold ions from solution [72]. SAMs and protein films formed on these surfaces are not always directly comparable to the much smoother surfaces described above.

Finally, for photobioelectrochemical applications, transparent surfaces can be beneficial. In this case, very thin gold films of $\leq$30 nm can be prepared by thermal evaporation or sputtering and thus also via a template stripping procedures. If the films are made too thin ($<<$30 nm), they might not form a continuous conducting gold layer unless particular care is taken in forming the layers. Although a thickness of 30 nm is a good compromise, such surfaces are only semi-transparent.

3.4 Gold Nanoparticles

Similar to carbon nanotubes and graphene nanoflakes, gold nanoparticles have been used to modify electrodes, be it bulk gold electrodes or other materials. The immediate benefit is the ability of these particles to enlarge enormously the surface area of the electrode, increasing the interfacial area to which proteins can interface. Second, the nanoparticles might be able to access otherwise occluded pockets on the surface of a redox protein.

As with macroscopic gold electrodes, gold nanoparticles can be modified with SAMs to provide a more 'biofriendly' surface for the proteins to interact with. For instance, 15-nm gold nanoparticles modified with a mixture of mercaptohexanol and hexanethiol enabled the electrochemical characterisation of three different membrane proteins (heme-copper oxidases) [73]. Importantly, macroscopic gold surfaces, even when modified with the same SAMs, do not give rise to an electrochemical response with the same proteins. This suggests that not only do the nanoparticles increase the surface area of the electrode, but their size is an important feature required for efficient electron transfer to these proteins. Similarly, gold nanoparticles with 3-mercapto-1-propanesulfonic acid were used to enhance the current of a photoelectrode containing PSI [74]. More elaborate strategies have also been reported. For instance, the modification of gold nanoparticles with a co-factor of glucose oxidase has been used specifically to couple this enzyme to the particles, resulting in an increased electronic coupling [75].

4 Transparent Electrodes (Indium Tin Oxide)

In the previous paragraphs it was highlighted that certain electrode materials, such as graphene or gold, can be fabricated as thin films. These thin films are specifically suitable for photobioelectrochemical applications as they are transparent. For macroscopic planar electrodes, transparency is not an absolute requirement in photoelectrochemistry, as the light-harvesting catalyst on the surface can be illuminated even if the electrode is non-transparent. Nonetheless, for certain applications, e.g. spectroelectrochemistry, transparent electrodes have considerable experimental benefits. Furthermore, it is advantageous to use porous or nanostructured electrode materials as the increased surface area leads to enhanced 'loading' of enzyme or catalyst on the surface and thus a higher current output (see Sect. 5). In this case, the electrode material has to be transparent for efficient illumination of enzymes on the surface. Porous or nanostructured electrodes with thin films of materials such as gold scatter light significantly, reducing the overall transparency. In this case, thin film coatings of transparent materials such as indium tin oxide, fluorine doped oxide or doped zinc oxide could be used. Indium tin oxide has been the most intensively used and is the industrial standard because of its high transparency and good conductive properties.

ITO has a negatively charged surface and, as with graphite, positively-charged polyelectrolytes such as polylysines can be used to improve interaction with negatively charged proteins. For instance, polylysine-modified ITO has been shown to enhance interaction with spinach ferridoxin in solution, giving rise to near-ideal solution cyclic voltammetry [76]. In contrast, positively-charged cytochrome *c* from yeast adsorbs directly on ITO [77]. For photobioelectrochemical application, both PSI and PSII have been immobilised on ITO [78, 79]. Although much less work has been performed with ITO than with carbon or gold electrodes, very similar strategies have been reported to improve (photo)bioelectrochemical responses. As the strategies are very similar to those described above, we do not give an in-depth description, but some typical examples reported with photosystems are briefly summarised here.

As already mentioned, ITO has been non-covalently modified with polyelectrolytes, such as polylysine [76]. An interesting approach was also given by non-covalent modification with a poly-benzylviologen [79], which is not only a polycation but also a redox-active polymer (see Sect. 6). Similar to the formation of SAMs on gold electrodes, phosphonic acids can be used to form self-assembled monolayers on ITO [78]. Covalent modification of ITO electrodes is also possible using silane chemistry, similar to that frequently performed on glass and silicon oxide surfaces. For instance, treatment with cyanopropyl triethoxysilane results in a strongly negatively charged surface with carboxylic acid groups [79]. ITO nanoparticles have also been used to increase the electrochemical surface area and to improve the electrochemical coupling to PSII [78]. Finally, PSII has been covalently attached to ITO surfaces using a commonly used coupling reagent

(EDC/NHS) to create a peptide bond between the amine groups on PSII and carboxylic acids on ITO [80].

5 Meso- and Macroporous Electrodes

In the previous sections, the use of nanoparticles, nanotubes, nanosheets and other 'nanoobjects' was introduced as a method to increase the electrochemical surface area of electrodes, which increases the amount of enzymes that can absorb and/or interact with the electrode. These nanoobjects have sizes that are of the same of order of magnitude as enzymes. The sizes of pores formed when depositing nanoobjects on an electrode surface are therefore too small for enzymes to populate. Unless the pores in this matrix are specifically engineered to be larger than the nanoobjects, only the very top interface between the nanoobjects and the solution is available to biomolecules. In principle it would be possible to mix the nanoobjects with the enzyme before applying the mixed particle/enzyme preparation to the electrode surface, but relatively few examples of this approach have appeared in the literature (for two examples see [81, 82]). Layered structures can be prepared where enzymes and nanoparticles are applied consecutively in several iterations, but this approach is relatively rarely described in the literature [83].

The alternative is to engineer the surface to contain meso- and macropores. Applications of mesoporous materials for enzyme immobilisation [84] and electrochemistry [85] have long been realised, although applications in protein bioelectrochemistry are still relatively scarce (see below). Two methods are available when engineering mesoporous electrode surfaces, top-down and bottom-up. In the top-down approach, one starts with a bulk macroscopically planar electrode and removes part of the surface to fabricate pores. In the bottom-up approach, meso-scale objects (e.g. nanotubes or nanospheres, see Fig. 5c, d, respectively) are deposited onto the planar electrode to engineer pores. A common variation of the latter approach is to deposit a mixture of two materials, one of which is a sacrificial material later removed to create a porous structure. Simply, if the structures shown in Fig. 5c, d are used as sacrificial templates that are 'backfilled', the nanotubular and inverse opal structures shown in Fig. 5a, b, respectively, can be generated. Note that for the inverse opal structures it is important to have sizable pores between the hollow spheres to make the porous material continuous for the enzymes to enter the mesoporous layer. Bottom-up, template-structured electrodes can generate much more exotic structures with higher surface areas than top-down engineered electrodes. Initially, templated materials were silica-based, which is not suitable for (bio)electrochemicals applications, but conducting mesoporous materials have now been prepared, including ordered porous metals [86], mesoporous carbons [87] and semi-conducting metal oxides [88, 89].

Depending on how the mesoporous electrodes are synthesized or how templating is performed, either ordered or disordered mesoporous electrodes can be formed with pore sizes ranging from two to several hundreds of nanometres. For enzyme

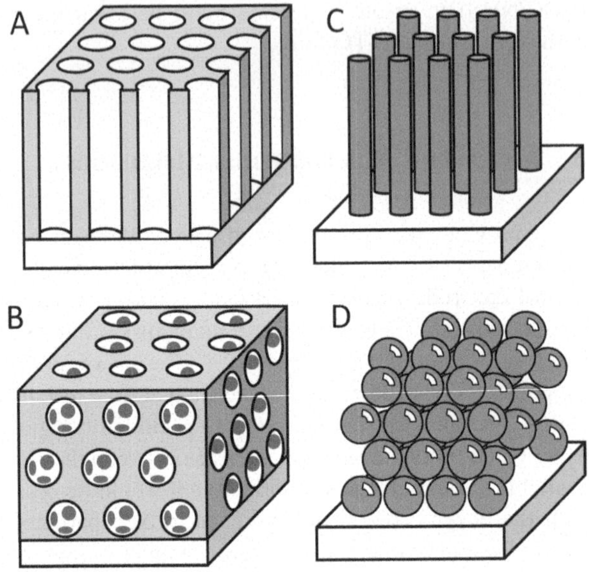

Fig. 5 Schematic representation of common mesoporous material architectures. (**a**) Tubes. (**b**) Inverse opal. (**c**) Wires. (**d**) Particles or spheres

immobilisation, sizes above 5–10 nm are generally required. Besides ordered and disordered, a distinction can be made between non-hierarchical and hierarchical structured electrodes. Non-hierarchical structured electrodes have a single type of geometry, typically either nanowires (or rods), nanotubes, nanoparticles or inverse opals and examples are schematically represented in Fig. 5. Hierarchical structures can be formed by employing multiple geometries on different length scales, e.g. adsorbing nanoparticles on surfaces containing microscale or nanoscale wires, creating a geometry of rough nano- or microwires.

The application of mesoporous materials to protein electrochemistry is still a relatively new field and the majority of the work has been performed with metal oxide electrodes, either nanotubes (Fig. 5a) or inverse opal electrodes (Fig. 5b). One of the first reports, which was published in 2003, adsorbed a model redox protein, cytochrome *c*, in an ordered mesoporous niobium oxide (MNO) film, which forms a 2D hollow tube material (Fig. 5a) [90]. The MNO was prepared by synthesizing niobium oxide on a surface in the presence of an amphiphile (a triblock polymer) which forms vertical ordered tubes. After removal of the amphiphile by baking (and calcination) at high temperature, an ordered MNO structure remains. It was found that tubes with a pore-diameter of 2 nm adsorbed only 1.8 μmol cytochrome *c*/g MNO, which is similar to the adsorption of cytochrome *c* to non-mesoporous niobium oxide films. This suggested that cytochrome *c* only adsorbs on the top surface of the MNO and not in the nanotubes. This is likely caused by the pore size, which is too small for cytochrome *c* to penetrate the hollow tubes. Increasing the pore size to 6 nm increased the adsorption to 2.5 μmol/g, which is an improvement, but still not as much as expected. It was hypothesized that cytochrome *c* 'plugs' the hollow tubes, limiting the ability of the MNO structure to take up protein.

In the same year, electrochemistry of cytochrome *c* on mesoporous titanium oxide electrodes was reported [83]. This film was made by a layer-by-layer deposition of cytochrome *c* and TiO_2 nanoparticles of 6 nm. The resulting structure formed discontinuous TiO_2 layers, which were not conductive, giving rise to diffusion-controlled electrochemistry of cytochrome *c* moving between the multilayers. Later, in 2007, is was shown that, if cytochrome *c* is adsorbed within a premade disordered mesoporous TiO_2 or SnO_2 film, the expected (non-diffusion) electrochemical response for an immobilised protein film is obtained [91]. In this case, polylysine was used to help immobilisation.

In 2008, haemoglobin was absorbed in a highly-ordered inverse opal mesoporous TiO_2 structure with a uniform pore size of 6 nm (Fig. 5b) [92]. The protein surface coverage (i.e. loading) was 139 pmol/cm^2, more than seven times the maximum monolayer coverage of a planar electrode. The great benefit of using SnO_2 mesoporous electrodes is that it is optically transparent and thus suitable for photobioelectrochemical systems and spectroelectrochemical characterization of, for instance, haeme proteins [93]. Besides SnO_2, transparent mesoporous ITO has also been reported [94]. Recently, a hierarchically structured ITO electrode was reported for the immobilisation of PSII and a hydrogenase in a photobiological water splitting device [89]. The hierarchically structured ITO electrode consisted of an inverse opal disordered structure with hollow spheres up to 750 nm wide and with 150 nm channels connecting the hollow spheres. The ITO material itself was made from fused ITO nanoparticles ($<$50 nm), resulting in a rough disorder porous ITO material lining the pores, thereby creating the hierarchical porous structure. Films up to 80 μm thick could be produced with protein loadings that were orders of magnitude larger than comparable 'planar' nanostructure ITO electrodes.

Disordered vertical nanowires (Fig. 5c) can be synthesized via controlled growth on seeded surfaces, where the 'seeds' catalyse the 1D growth of wires away from the electrode. The spacing between the wires is not easily controlled, but is often much larger than the enzyme diameter and thus suitable. Glucose oxidase was shown to exhibit direct electron transfer with carbon-decorated ZnO nanowires, although it was not shown whether the enzyme infiltrated the wires or adsorbed on top of the wires [95]. So-called nanowire forests have also been reported for ITO [96].

6 Encapsulation Redox Proteins in Redox-Active Polymers

Although the use of mediators has been mentioned, the focus of this chapter so far has been on direct interaction between the redox protein and the electrode surface. Some examples of diffusion-controlled protein electrochemistry have been given, but the majority of the electrodes covered in this chapter use a layer (or film) of immobilised proteins. Different methods to adsorb the proteins on the electrode have been covered, most of them relying on direct interaction between the proteins and the conducting electrode material. There are, however, two other widely used

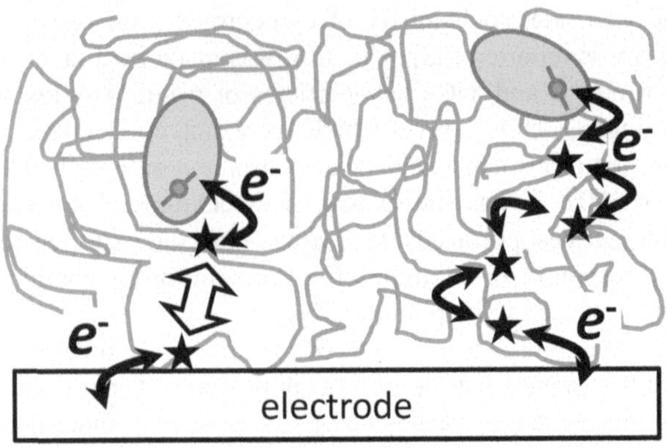

Fig. 6 Schematic representation of proteins encapsulated in hydrogels where the electrons are transferred to the protein via (*left*) encapsulated free-diffusion mediators or (*right*) mediators covalently bound to the polymer to create a redox-active hydrogel. The *solid arrows* represent electron transport and the *open arrow* represents diffusion of the mediators

methods to entrap a protein on an electrode surface. The first uses semi-permeable membranes which trap proteins in a small volume close to the electrode surface. This method does not directly involve the selection or modification of the electrode material itself and is not discussed further here. The second method is the encapsulation of the redox enzyme into polymer hydrogels on the surface. Encapsulation of proteins in hydrogels is a method that precedes protein electrochemistry and has many applications outside bioelectrochemistry [97]. Besides hydrogels, proteins have been encapsulated in sol–gels, usually silicates [98].

Hydrogels distinctly differ from the use of polylysine described in Sect. 2.2.1 Despite the fact that polylysine is a polymer, it forms a thin layer on the surface which is not thick enough to encapsulate proteins (unless multilayered systems are formed). For the encapsulation of enzymes into a hydrogel or sol–gel, either the protein is mixed with a hydrogel before casting them on the surface as a micron-thick film or the polymerisation is performed in the presence of the protein. Because hydrogels and sol–gels are typically microns thick and diffusion of enzyme in these matrices is slow or non-existent, electron transfer between the electrode surface and enzyme has to be mediated.

There are broadly two methods by which electron transfer can be mediated through hydrogels (Fig. 6). The first is the addition of small soluble mediators to the hydrogel. These small molecules diffuse through the gel and can shuttle between enzyme and electrode on a sufficiently fast timescale. Because the mediators are free to diffuse within the hydrogel, they typically leach out of the hydrogel and thus the mediator has to be added to the electrolyte. Assuming the redox reactions between electrode, mediator and enzymes are fast, the electron transfer rate is limited by the concentration and diffusion kinetics of the mediator in the hydrogel. Furthermore, the thermodynamics (i.e. the potential) of the system is controlled mainly by the redox potential of the mediator and not that of the enzyme.

The second method is to bind the mediators covalently to the hydrogel polymer or, in the case of inorganic mediators such as osmium, via coordinate bonds to ligands, which in turn are covalently bound to the hydrogel [99]. Redox-active hydrogels commonly use osmium complexes, quinones or viologens as mediators. As these mediators are bound to the hydrogel they have limited mobility. This means the mediators cannot leak out of the hydrogel, but neither is it possible for the mediators to diffuse between the electrode and redox proteins that are encapsulated. It is thus necessary to incorporate enough of the mediator in the hydrogel so that electrons can hop from one mediator to the next. The kinetics of electron exchange between the enzyme and electrode could be limited by the electron hopping between mediators, but the rate-limiting step is not always easy to determine as it is not always possible to quantify unambiguously the enzyme concentration in the hydrogel.

One of the properties that can be optimised in 'mediated' systems (other than the mediator concentration) is the redox potential of the mediators. This needs to lie as close as possible to the redox potential of the protein, reducing the energy loss of electron transfer from electrode to mediator to enzyme (or vice versa).

Redox-active polymers are, in effect, conductive, but they should not be confused with what is also known as 'conducting polymers' such as doped polyaniline, which are coined conducting because they contain delocalised π electrons [100]. Although these polymers are indeed more conductive than 'non-conducting' organic media or hydrogels, they are generally not conductive enough for enzyme-based fuel cells or photobioelectrochemical systems (i.e. the conducting polymers cannot raise the electronic coupling between the enzyme and the electrode high enough to enable sub-second electron transfer rates). This is in contrast to the redox-active polymers, where electron transfer rates exceeding 100 s^{-1} are reported.

Several groups have studied whether the conductivity of redox-active hydrogels or polymers can be further improved by the incorporation of (semi)-conductive nanoparticles (for an example see [101]). In some cases, the nanoparticles themselves are redox-active, thus contributing the 'redox-active' element of the polymer [102].

Redox-active polymers have also been used in photobioelectrochemical applications and this is more extensively discussed in Chap. 4 of this book. In brief, both PSI and PSII have been encapsulated in hydrogels [79, 103–108]. The use of free mediators has been reported [105] as well as the used of redox-active polymers [79, 103, 104, 106–108]. Because of the favourable reduction potential and the fact that the reduction potential can be tuned by altering the ligands, osmium complexes are the mediator of choice [103–109]. However, osmium mediators with optimal reduction potentials for PSII can interfere by reducing oxygen, and alternative compounds such as phenothiazine and viologen have also been used [79, 106]. Besides purified photosystems, redox-active polymers have been used to couple complete thylakoid membranes and cyanobacteria to electrodes [109, 110].

References

1. Eddowes MJ, Hill HAO (1977) Novel method for investigation of electrochemistry of metalloproteins – cytochrome *c*. J Chem Soc Chem Commun 771–772
2. Zhang J, Chi Q, Dong S, Wang E (1996) In situ electrochemical scanning tunnelling microscopy investigation of structure for horseradish peroxidase and its electrocatalytic property. Bioelectrochem Bioenerg 39:267–274
3. Zhang J, Chi Q, Zhang B, Dong S, Wang E (1998) Molecular characterization of beef liver catalase by scanning tunneling microscopy. Electroanalysis 10:738–746
4. Arrigan DWM, Bartlett PN (1998) A scanning force microscopy study of poly(phenol) films containing immobilized glucose oxidase. Biosens Bioelectron 13:293–304
5. Lukins PB, Barton CS (2003) Evidence for spatially-coherent trans-molecular electron tunnelling through two-dimensional arrays of photosystem II core complexes. Chem Commun 602–603
6. Rusling JF (1998) Enzyme bioelectrochemistry in cast biomembrane-like films. Acc Chem Res 31:363–369
7. de Groot MT, Merkx M, Koper MTM (2005) Heme release in myoglobin-DDAB films and its role in electrochemical NO reduction. J Am Chem Soc 127:16224–16232
8. de Groot MT, Merkx M, Koper MTM (2007) Evidence for heme release in layer-by-layer assemblies of myoglobin and polystyrenesulfonate on pyrolitic graphite. J Biol Inorg Chem 12:761–766
9. Krishnan S, Schenkman JB, Rusling JF (2011) Bioelectronic delivery of electrons to cytochrome P450 enzymes. J Phys Chem B 115:8371–8380
10. Armstrong FA, Bond AM, Hill HAO, Oliver BN, Psalti ISM (1989) Electrochemistry of cytochrome *c* plastocyanin and ferredoxin at edge and basal plane graphite electrodes interpreted via a model based on electron transfer at electroactive sites of microscopic dimensions in size. J Am Chem Soc 111:9185–9189
11. Blanford CF, Armstrong FA (2006) The pyrolytic graphite surface as an enzyme substrate: microscopic and spectroscopic studies. J Solid State Electrochem 10:826–832
12. Barker PD, Mauk AG (1992) pH-linked conformational regulation of a metalloprotein oxidation reduction equilibrium - electrochemical analysis of the alkaline form of cytochrome *c*. J Am Chem Soc 114:3619–3624
13. Avila A, Gregory BW, Niki K, Cotton TM (2000) An electrochemical approach to investigate gated electron transfer using a physiological model system: cytochrome *c* immobilized on carboxylic acid-terminated alkanethiol self-assembled monolayers on gold electrodes. J Phys Chem B 104:2759–2766
14. Jeuken LJC, Armstrong FA (2001) Electrochemical origin of hysteresis in the electron-transfer reactions of adsorbed proteins: contrasting behavior of the "blue" copper protein, azurin, adsorbed on pyrolytic graphite and modified gold electrodes. J Phys Chem B 105:5271–5282
15. Harris PJF (2004) Fullerene-related structure of commercial glassy carbons. Philos Mag 84:3159–3167
16. Hagen WR (1989) Direct electron transfer of redox proteins at the bare glassy carbon electrode. Eur J Biochem 182:523–530
17. Heering HA, Bulsink YBM, Hagen WR, Meyer TE (1995) Influence of charge and polarity on the redox potentials of high-potential iron-sulfur proteins: evidence for the existence of two groups. Biochemistry 34:14675–14686
18. Johnson DL, Maxwell CJ, Losic D, Shapter JG, Martin LL (2002) The influence of promoter and of electrode material on the cyclic voltammetry of *Pisum sativum* plastocyanin. Bioelectrochemistry 58:137–147
19. Butt JN et al (1991) Investigation of metal-ion uptake reactivities of 3Fe-4S clusters in proteins - voltammetry of coadsorbed ferredoxin aminocyclitol films at graphite-electrodes and spectroscopic identification of transformed clusters. J Am Chem Soc 113:6663–6670

20. Butt JN et al (1993) Voltammetric characterization of rapid and reversible binding of an exogenous thiolate ligand at a 4Fe-4S cluster in ferredoxin-III from *Desulfovibrio africanus*. J Am Chem Soc 115:1413–1421
21. Rapson TD, Kappler U, Bernhardt PV (2008) Direct catalytic electrochemistry of sulfite dehydrogenase: mechanistic insights and contrasts with related Mo enzymes. Biochim Biophys Acta Bioenerg 1777:1319–1325
22. Forzani ES, Teijelo ML, Nart F, Calvo EJ, Solis VM (2003) Effect of the polycation nature on the structure of layer-by-layer electrostatically self-assembled multilayers of polyphenol oxidase. Biomacromolecules 4:869–879
23. Zhou Y, Hu N, Zeng Y, Rusling JF (2002) Heme protein-clay films: direct electrochemistry and electrochemical catalysis. Langmuir 18:211–219
24. Zhang Y, Magdaong NM, Shen M, Frank HA, Rusling JF (2015) Efficient photoelectrochemical energy conversion using spinach photosystem II (PSII) in lipid multilayer films. ChemistryOpen 4:111–114
25. Pinson J, Podvorica F (2005) Attachment of organic layers to conductive or semiconductive surfaces by reduction of diazonium salts. Chem Soc Rev 34:429–439
26. Combellas C, Kanoufi F, Pinson J, Podvorica FI (2005) Time-of-flight secondary ion mass spectroscopy characterization of the covalent bonding between a carbon surface and aryl groups. Langmuir 21:280–286
27. Combellas C, Jiang DE, Kanoufi F, Pinson J, Podvorica FI (2009) Steric effects in the reaction of aryl radicals on surfaces. Langmuir 25:286–293
28. Combellas C, Kanoufi F, Pinson J, Podvorica FI (2008) Sterically hindered diazonium salts for the grafting of a monolayer on metals. J Am Chem Soc 130:8576–8577
29. Clausmeyer J, Henig J, Schuhmann W, Plumere N (2014) Scanning droplet cell for chemoselective patterning through local electroactivation of protected quinone monolayers. ChemPhysChem 15:151–156
30. Leroux YR, Fei H, Noel JM, Roux C, Hapiot P (2010) Efficient covalent modification of a carbon surface: use of a silyl protecting group to form an active monolayer. J Am Chem Soc 132:14039–14041
31. Guo K et al (2013) Effects of surface charge and hydrophobicity on anodic biofilm formation, community composition, and current generation in bioelectrochemical systems. Environ Sci Tech 47:7563–7570
32. Picot M, Lapinsonniere L, Rothballer M, Barriere F (2011) Graphite anode surface modification with controlled reduction of specific aryl diazonium salts for improved microbial fuel cells power output. Biosens Bioelectron 28:181–188
33. Sosna M, Chretien JM, Kilburn JD, Bartlett PN (2010) Monolayer anthracene and anthraquinone modified electrodes as platforms for *Trametes hirsuta* laccase immobilisation. Phys Chem Chem Phys 12:10018–10026
34. Blanford CF, Heath RS, Armstrong FA (2007) A stable electrode for high-potential, electrocatalytic O_2 reduction based on rational attachment of a blue copper oxidase to a graphite surface. Chem Commun 1710–1712
35. Davis JJ, Coles RJ, Hill HAO (1997) Protein electrochemistry at carbon nanotube electrodes. J Electroanal Chem 440:279–282
36. Wooten M, Karra S, Zhang M, Gorski W (2014) On the direct electron transfer, sensing, and enzyme activity in the glucose oxidase/carbon nanotubes system. Anal Chem 86:752–757
37. Yan Y et al (2005) Bioelectrochemically functional nanohybrids through co-assembling of proteins and surfactants onto carbon nanotubes: facilitated electron transfer of assembled proteins with enhanced faradic response. Langmuir 21:6560–6566
38. Wang J, Li M, Shi Z, Li N, Gu Z (2002) Direct electrochemistry of cytochrome *c* at a glassy carbon electrode modified with single-wall carbon nanotubes. Anal Chem 74:1993–1997
39. Kim JH, Jin JH, Lee JY, Park EJ, Min NK (2012) Covalent attachment of biomacromolecules to plasma-patterned and functionalized carbon nanotube-based devices for electrochemical biosensing. Bioconjug Chem 23:2078–2086

40. Wu Y, Hu S (2007) Direct electron transfer of xanthine oxidase and its catalytic reduction to nitrate. Anal Chim Acta 602:181–186
41. Walgama C, Means N, Materer NF, Krishnan S (2015) Edge-to-edge interaction between carbon nanotube-pyrene complexes and electrodes for biosensing and electrocatalytic applications. Phys Chem Chem Phys 17:4025–4028
42. Gobel G, Lisdat F (2008) Organic interlayers for oxygen reducing electrodes based on bilirubin oxidase and MWCNT modified gold. Electrochem Commun 10:1691–1694
43. Giroud F, Minteer SD (2013) Anthracene-modified pyrenes immobilized on carbon nanotubes for direct electroreduction of O_2 by laccase. Electrochem Commun 34:157–160
44. van der Felt C et al (2011) Electron-transfer rates govern product distribution in electrochemically-driven P450-catalyzed dioxygen reduction. J Inorg Biochem 105:1350–1353
45. Tasca F, Harreither W, Ludwig R, Gooding JJ, Gorton L (2011) Cellobiose dehydrogenase aryl diazoniunn modified single walled carbon nanotubes: enhanced direct electron transfer through a positively charged surface. Anal Chem 83:3042–3049
46. Sekar N, Umasankar Y, Ramasamy RP (2014) Photocurrent generation by immobilized cyanobacteria via direct electron transport in photo-bioelectrochemical cells. Phys Chem Chem Phys 16:7862–7871
47. Zuo X et al (2010) Graphene oxide-facilitated electron transfer of metalloproteins at electrode surfaces. Langmuir 26:1936–1939
48. Patolsky F, Tao G, Katz E, Willner I (1998) C_{60}-mediated bioelectrocatalyzed oxidation of glucose with glucose oxidase. J Electroanal Chem 454:9–13
49. Feifel SC, Stieger KR, Lokstein H, Lux H, Lisdat F (2015) High photocurrent generation by photosystem I on artificial interfaces composed of pi-system-modified graphene. J Mater Chem A 3:12188–12196
50. Baffert C et al (2012) Covalent attachment of FeFe hydrogenases to carbon electrodes for direct electron transfer. Anal Chem 84:7999–8005
51. Rudiger O, Abad JM, Hatchikian EC, Fernandez VM, De Lacey AL (2005) Oriented immobilization of *Desulfovibrio gigas* hydrogenase onto carbon electrodes by covalent bonds for nonmediated oxidation of H_2. J Am Chem Soc 127:16008–16009
52. Bourdillon C et al (1992) Immobilization of glucose-oxidase on a carbon surface derivatized by electrochemical reduction of diazonium salts. J Electroanal Chem 336:113–123
53. Polsky R et al (2007) Diazonium-functionalized horseradish peroxidase immobilized via addressable electrodeposition: direct electron transfer and electrochemical detection. Langmuir 23:364–366
54. Love JC, Estroff LA, Kriebel JK, Nuzzo RG, Whitesides GM (2005) Self-assembled monolayers of thiolates on metals as a form of nanotechnology. Chem Rev 105:1103–1169
55. Song S, Clark RA, Bowden EF, Tarlov MJ (1993) Characterization of cytochrome *c*/alkanethiolate structures prepared by self-assembly on gold. J Phys Chem 97:6564–6572
56. Arnold S, Feng ZQ, Kakiuchi T, Knoll W, Niki K (1997) Investigation of the electrode reaction of cytochrome *c* through mixed self-assembled monolayers of alkanethiols on gold (111) surfaces. J Electroanal Chem 438:91–97
57. Jeuken LJC, McEvoy JP, Armstrong FA (2002) Insights into gated electron-transfer kinetics at the electrode-protein interface: a square wave voltammetry study of the blue copper protein azurin. J Phys Chem B 106:2304–2313
58. Chi QC, Zhang J, Andersen JET, Ulstrup J (2001) Ordered assembly and controlled electron transfer of the blue copper protein azurin at gold(111) single-crystal substrates. J Phys Chem B 105:4669–4679
59. Krzeminski L et al (2011) Orientational control over nitrite reductase on modified gold electrode and its effects on the interfacial electron transfer. J Phys Chem B 115:12607–12614
60. Hwang ET et al (2015) A decaheme cytochrome as a molecular electron conduit in dye-sensitized photoanodes. Adv Funct Mater 25:2308–2315

61. Uehara TM, de Aguiar HB, Bergamaski K, Miranda PB (2014) Adsorption of alkylthiol self-assembled monolayers on gold and the effect of substrate roughness: a comparative study using scanning tunneling microscopy, cyclic voltammetry, second-harmonic generation, and sum-frequency generation. J Phys Chem C 118:20374–20382
62. Chi Q et al (2000) Molecular monolayers and interfacial electron transfer of *Pseudomonas aeruginosa* azurin on Au(111). J Am Chem Soc 122:4047–4055
63. Heering HA, Wiertz FGM, Dekker C, de Vries S (2004) Direct immobilization of native yeast iso-1 cytochrome *c* on bare gold: fast electron relay to redox enzymes and zeptomole protein-film voltammetry. J Am Chem Soc 126:11103–11112
64. Hasan MN, Kwakernaak C, Sloof WG, Hagen WR, Heering HA (2006) *Pyrococcus furiosus* 4Fe-ferredoxin, chemisorbed on gold, exhibits gated reduction and ionic strength dependent dimerization. J Biol Inorg Chem 11:651–662
65. Friedrich MG, Robertson JWF, Walz D, Knoll W, Naumann RLC (2008) Electronic wiring of a multi-redox site membrane protein in a biomimetic surface architecture. Biophys J 94:3698–3705
66. Tel-Vered R, Yildiz HB, Yan YM, Willner I (2010) Plugging into enzymes with light: photonic "wiring" of enzymes with electrodes for photobiofuel cells. Small 6:1593–1597
67. Hess CR et al (2003) Gold electrodes wired for coupling with the deeply buried active site of *Arthrobacter globiformis* amine oxidase. J Am Chem Soc 125:7156–7157
68. Terasaki N et al (2009) Plugging a molecular wire into photosystem I: reconstitution of the photoelectric conversion system on a gold electrode. Angew Chem Int Ed 48:1585–1587
69. Fischer LM et al (2009) Gold cleaning methods for electrochemical detection applications. Microelectron Eng 86:1282–1285
70. Li ZG et al (2014) Effect of surface pretreatment on self-assembly of thiol-modified DNA monolayers on gold electrode. J Electroanal Chem 722:131–140
71. Stamou D et al (1997) Uniformly flat gold surfaces: imaging the domain structure of organic monolayers using scanning force microscopy. Langmuir 13:2425–2428
72. Miyake H, Ye S, Osawa M (2002) Electroless deposition of gold thin films on silicon for surface-enhanced infrared spectroelectrochemistry. Electrochem Commun 4:973–977
73. Meyer T et al (2014) Evidence for distinct electron transfer processes in terminal oxidases from different origin by means of protein film voltammetry. J Am Chem Soc 136:10854–10857
74. Terasaki N et al (2006) Fabrication of novel photosystem I-gold nanoparticle hybrids and their photocurrent enhancement. Thin Solid Films 499:153–156
75. Xiao Y, Patolsky F, Katz E, Hainfeld JF, Willner I (2003) "Plugging into enzymes": nanowiring of redox enzymes by a gold nanoparticle. Science 299:1877–1881
76. Wirtz M, Klucik J, Rivera M (2000) Ferredoxin-mediated electrocatalytic dehalogenation of haloalkanes by cytochrome P450(cam). J Am Chem Soc 122:1047–1056
77. El Kasmi A et al (2002) Adsorptive immobilization of cytochrome *c* on indium/tin oxide (ITO): electrochemical evidence for electron transfer-induced conformational changes. Electrochem Commun 4:177–181
78. Kato M, Cardona T, Rutherford AW, Reisner E (2012) Photoelectrochemical water oxidation with photosystem II integrated in a mesoporous, indium tin oxide electrode. J Am Chem Soc 134:8332–8335
79. Yehezkeli O, Tel-Vered R, Michaeli D, Nechushtai R, Willner I (2013) Photosystem I (PSI)/photosystem II (PSII)-based photo-bioelectrochemical cells revealing directional generation of photocurrents. Small 9:2970–2978
80. Kato M, Cardona T, Rutherford AW, Reisner E (2013) Covalent immobilization of oriented photosystem II on a nanostructured electrode for solar water oxidation. J Am Chem Soc 135:10610–10613
81. Quinson J et al (2014) Comparison of carbon materials as electrodes for enzyme electrocatalysis: hydrogenase as a case study. Faraday Discuss 172:473–496

82. Sun W, Zhai ZQ, Jiao K (2008) Hemoglobin modified carbon paste electrode: direct electrochemistry and electrocatalysis. Anal Lett 41:2819–2831
83. McKenzie KJ, Marken F (2003) Accumulation and reactivity of the redox protein cytochrome *c* in mesoporous films of TiO_2 phytate. Langmuir 19:4327–4331
84. Zhou Z, Hartmann M (2013) Progress in enzyme immobilization in ordered mesoporous materials and related applications. Chem Soc Rev 42:3894–3912
85. Walcarius A (2013) Mesoporous materials and electrochemistry. Chem Soc Rev 42:4098–4140
86. Lu L, Eychmuller A (2008) Ordered macroporous bimetallic nanostructures: design, characterization, and applications. Acc Chem Res 41:244–253
87. Liang CD, Li ZJ, Dai S (2008) Mesoporous carbon materials: synthesis and modification. Angew Chem Int Ed 47:3696–3717
88. Ren Y, Ma Z, Bruce PG (2012) Ordered mesoporous metal oxides: synthesis and applications. Chem Soc Rev 41:4909–4927
89. Mersch D et al (2015) Wiring of photosystem II to hydrogenase for photoelectrochemical water splitting. J Am Chem Soc 137:8541–8549
90. Xu X et al (2003) Ordered mesoporous niobium oxide film: a novel matrix for assembling functional proteins for bioelectrochemical applications. Adv Mater 15:1932–1936
91. Topoglidis E et al (2005) Immobilization and electrochemistry of negatively charged proteins on modified nanocrystalline metal oxide electrodes. Electroanalysis 17:1035–1041
92. Jia NQ et al (2008) Direct electrochemistry and enzymatic activity of hemoglobin immobilized in ordered mesoporous titanium oxide matrix. Electrochem Commun 10:774–777
93. Marritt SJ et al (2008) Spectroelectrochemical characterization of a pentaheme cytochrome in solution and as electrocatalytically active films on nanocrystalline metal-oxide electrodes. J Am Chem Soc 130:8588–8589
94. Aksu Y, Frasca S, Wollenberger U, Driess M, Thomas A (2011) A molecular precursor approach to tunable porous tin-rich indium tin oxide with durable high electrical conductivity for bioelectronic devices. Chem Mater 23:1798–1804
95. Liu JP et al (2009) Carbon-decorated ZnO nanowire array: a novel platform for direct electrochemistry of enzymes and biosensing applications. Electrochem Commun 11:202–205
96. Li LP et al (2015) Controlled synthesis of tin-doped indium oxide (ITO) nanowires. J Cryst Growth 413:31–36
97. Scouten WH (1987) A survey of enzyme coupling techniques. Methods Enzymol 135:30–65
98. Jin W, Brennan JD (2002) Properties and applications of proteins encapsulated within sol–gel derived materials. Anal Chim Acta 461:1–36
99. Willner I, Yan YM, Willner B, Tel-Vered R (2009) Integrated enzyme-based biofuel cells-a review. Fuel Cells 9:7–24
100. Dhand C, Das M, Datta M, Malhotra BD (2011) Recent advances in polyaniline based biosensors. Biosens Bioelectron 26:2811–2821
101. Joshi PP, Merchant SA, Wang YD, Schmidtke DW (2005) Amperometric biosensors based on redox polymer-carbon nanotube-enzyme composites. Anal Chem 77:3183–3188
102. Hicks JF, Zamborini FP, Osisek A, Murray RW (2001) The dynamics of electron self-exchange between nanoparticles. J Am Chem Soc 123:7048–7053
103. Badura A et al (2008) Photo-induced electron transfer between photosystem 2 via cross-linked redox hydrogels. Electroanalysis 20:1043–1047
104. Badura A et al (2011) Photocurrent generation by photosystem 1 integrated in crosslinked redox hydrogels. Energy Environ Sci 4:2435–2440
105. Baker DR, Simmerman RF, Sumner JJ, Bruce BD, Lundgren CA (2014) Photoelectrochemistry of photosystem I bound in nafion. Langmuir 30:13650–13655
106. Hartmann V et al (2014) Redox hydrogels with adjusted redox potential for improved efficiency in Z-scheme inspired biophotovoltaic cells. Phys Chem Chem Phys 16:11936–11941

107. Kothe T et al (2014) Engineered electron-transfer chain in photosystem 1 based photocathodes outperforms electron-transfer rates in natural photosynthesis. Chem Eur J 20:11029–11034
108. Zhao FY, Sliozberg K, Rogner M, Plumere N, Schuhmann W (2014) The role of hydrophobicity of Os-complex-modified polymers for photosystem 1 based photocathodes. J Electrochem Soc 161:H3035–H3041
109. Hamidi H et al (2015) Photocurrent generation from thylakoid membranes on osmium-redox-polymer-modified electrodes. ChemSusChem 8:990–993
110. Hasan K et al (2014) Photo-electrochemical communication between cyanobacteria (*Leptolyngbia* sp.) and osmium redox polymer modified electrodes. Phys Chem Chem Phys 16:24676–24680

Adv Biochem Eng Biotechnol (2017) 158: 75–110
DOI: 10.1007/10_2016_3

Published online: 31 July 2016

Vibrational Spectroscopic Techniques for Probing Bioelectrochemical Systems

Philip A. Ash and Kylie A. Vincent

Abstract A more complete understanding of bioelectrochemical interfaces is of increasing importance in both fundamental studies and biotechnological applications of proteins. Bioelectrochemical methods provide detailed information about the activity or rate of a process, but in situ spectroscopic methods are needed to gain direct structural insight into functionally relevant states. A number of methods have been reported that allow electrochemical and spectroscopic data to be collected from the same electrode, providing direct spectroscopic 'snapshots' of protein function, and here we focus on the application of infrared and Raman spectroscopies to the study of electrode-immobilised species. The ability to probe coordination at metal centres, protonation changes in amino acid side chains, reaction-induced changes in organic cofactors or substrates, protein orientation and subtle changes in protein secondary structure simultaneously, rapidly and at room temperature means that vibrational spectroscopic approaches are almost uniquely applicable to answering a wide range of questions in bioelectrochemistry.

Keywords Spectroelectrochemistry, In situ spectroscopy, Biocatalysis, Electrocatalysis, Infrared, Raman, SERS, SERRS, SEIRA, Membrane protein, Redox enzyme

Contents

P.A. Ash and K.A. Vincent (✉)
Department of Chemistry, Inorganic Chemistry Laboratory, University of Oxford, South Parks Road, Oxford OX1 3QR, UK
e-mail: kylie.vincent@chem.ox.ac.uk

1 Introduction

Vibrational spectroscopy has been used in a number of ways to study biological systems [1], and this chapter concerns the development and application of approaches that extend its scope by coupling spectroscopy with bioelectrochemistry. There have been many studies in which vibrational spectroscopy is applied to redox processes associated with proteins in solution, but here we focus on examples in which the biomolecule is immobilised on an electrode surface, thus making it possible to combine spectroscopy with the precise potential control provided by protein film electrochemistry [2, 3]. Vibrational spectroscopy is not limited to studying the protein itself but is also sensitive to the protein/electrode interface, meaning that it is possible to probe the nature and efficiency of different immobilisation strategies.

Infrared and Raman spectroscopies are the main techniques used to analyse the vibrational structure of biomolecules, and are amongst the spectroscopic techniques most easily coupled to the study of electrode surfaces. Although infrared (IR) and Raman spectroscopies both probe the vibrational energy levels of molecules, they have fundamentally different physical origins. Infrared spectroscopy is an absorption spectroscopy, and measures the excitation of molecular vibrations by direct absorption of IR radiation. Raman spectroscopy is based on inelastic scattering of monochromatic light (light of a single energy); a small proportion of the incident energy is transferred to the molecule in the form of vibration, and so the energy of the Raman scattered light is 'shifted' by an amount equivalent to the energy of the molecular vibration.

2 Electromagnetic Spectrum and Molecular Vibrations

Although much of modern spectroscopy is heavily reliant on quantum mechanical descriptions of molecules, matter and radiation (here the reader is directed to the more detailed texts of other authors) [4–6], the application of vibrational spectroscopy to biological systems can be conceptually understood within a classical framework. Light is a collection of electromagnetic waves, consisting of oscillating electric and magnetic fields, and the continuum of all energies of light is called the electromagnetic spectrum. The energy, E (in Joules, J), of a given wavelength, λ (in m), of radiation is given by

$$E = \frac{hc}{\lambda} = h\nu, \tag{1}$$

where h is Planck's constant, c is the speed of light (in m s^{-1}) and ν is the frequency of the radiation in Hz. The infrared region of the electromagnetic spectrum, in particular the so-called mid-infrared with wavelengths between roughly 2.5 and 100 μm, spans the energy range of molecular vibrations.

When a molecule vibrates, the periodic displacement of the atoms within that molecule leaves the centre of mass unchanged (displacement of the centre of mass is simply translation). We can therefore define a vibrational coordinate, q, which describes the vibrational motion of a molecule in terms of the displacement of each atom from its equilibrium position. For the simplest case of a diatomic molecule this can be expressed in terms of the instantaneous and equilibrium bond lengths, r and r_e respectively:

$$q = \Delta r = r - r_e. \tag{2}$$

This vibrational coordinate can be used to describe the potential energy associated with any given vibrational motion by solving the equation of motion for the electrons within the molecule with the positions of the nuclei fixed at various points along the vibrational coordinate (this takes advantage of the Born–Oppenheimer separation, which states that because electronic motion is so much faster than nuclear motion the two can be treated independently). As such, the potential well of a vibration gives us the sum of the kinetic energy of the electrons and the electrostatic energy of the nuclei at all points along the vibrational coordinate. In the simplest case, the potential well, U_{vib}, can be approximated with a parabolic function of the vibrational coordinate:

$$U_{\text{vib}} = \frac{1}{2}kq^2. \tag{3}$$

This corresponds to the classical simple harmonic oscillator: the restoring force is directly proportional to the vibrational coordinate ((4), where the negative sign reflects the fact that the restoring force always acts towards r_e) and the force constant k (i.e. bond strength) is independent of the position along the vibrational coordinate (5).

$$F_q = -\frac{dU_{\text{vib}}}{dq} = -kq, \tag{4}$$

$$\frac{d^2U_{\text{vib}}}{dq^2} = k. \tag{5}$$

The total energy, $E_{\text{vib, total}}$, of the harmonic oscillator includes a kinetic energy contribution (because the atoms within the molecule are necessarily in motion during a vibration).

$$E_{\text{vib, total}} = \frac{1}{2}\frac{p^2}{\mu} + \frac{1}{2}kq^2. \tag{6}$$

Here p is the instantaneous momentum of the molecule and μ is the effective mass. Because the total energy must be conserved, $E_{\text{vib, total}}$ is constant and its first derivative is equal to zero. This is equivalent to applying Newton's second law (force is the product of mass and acceleration) to a mass equal to μ, which leads to the equation of motion for the simple harmonic oscillator:

$$\frac{\partial^2 q}{\partial t^2} + \frac{k}{\mu}q = 0. \tag{7}$$

The equation of motion is an ordinary differential equation, a general solution of which can be written as a cosine function in a form that immediately allows identification of the amplitude, A, frequency in radians per second, ω, and phase angle, ϕ, of the vibration:

$$q = A\cos(\omega t + \phi). \tag{8}$$

Taking the second derivative of (8) allows the frequency of the vibration to be expressed as a function of the force constant and effective mass, by direct comparison with the equation of motion:

$$\frac{\partial^2 q}{\partial t^2} = -\omega^2 A\cos(\omega t + \phi) = -\omega^2 q, \tag{9}$$

$$\frac{\partial^2 q}{\partial t^2} + \omega^2 q = 0, \tag{10}$$

$$\omega = \sqrt{\frac{k}{\mu}}. \tag{11}$$

2.1 A Note on Relative Units and Energy

Equation (11) expresses the frequency of a molecular vibration in units of radians per second. The frequency is readily converted to units of Hz (to express the frequency in cycles per second) by dividing by the number of radians in a complete revolution (i.e. a circle, (12)), such that the energy of the vibration in Joules is given by (13).

$$\nu = \frac{1}{2\pi}\sqrt{\frac{k}{\mu}}, \tag{12}$$

$$E = \frac{hc}{\lambda} = h\nu = \frac{h}{2\pi}\sqrt{\frac{k}{\mu}}. \tag{13}$$

The measurement of physical quantities such as wavelength or frequency can be carried out more accurately than uncertainty in the value of Planck's constant, h. The inverse of the wavelength, $^1/_\lambda$,[1] is known as the wavenumber, $\widetilde{\nu}$, and provides a useful relative unit for the energy of a molecular vibration:

$$\widetilde{\nu} = \frac{E}{hc} = \frac{1}{2\pi c}\sqrt{\frac{k}{\mu}}. \tag{14}$$

Wavenumber, $\widetilde{\nu}$, is therefore an almost universally used relative energy unit in vibrational spectroscopy, in units of cm^{-1} (i.e. the speed of light in (14) is expressed in cm s^{-1}). As is apparent from (14), $\widetilde{\nu}$ is *proportional* to both energy and frequency: the terms 'wavenumber' and 'frequency' are often used interchangeably in the literature but this is technically not correct, although wavenumber units can be thought of as a 'spatial frequency' as they correspond to the number of cycles (wavelengths) per unit distance.

Typical molecular vibrations lie in the wavenumber range 100–4,000 cm^{-1}, corresponding to wavelengths of 100–2.5 μm, frequencies of ca. 2,998–119,920 GHz and energies of ca. 1.9864×10^{-21} to 7.9458×10^{-20} J or 1.2398×10^{-2} to 0.49594 eV (where an electron-volt, eV, is the amount of energy gained by an electron accelerated through a potential of 1 V). Conversion factors between wavenumbers and other common energy or relative energy units are listed in Table 1.

At all temperatures above absolute zero, molecules possess an amount of thermal energy, kT, where k is the Boltzman constant (1.38066×10^{23} J K^{-1}) and

Table 1 Conversion factors for direct comparison of wavenumber with other relative energy units

	1 cm^{-1}	1 GHz	1 eV	1 J	1 K
cm^{-1}	1	29.97925	1.2398×10^{-4}	1.9864×10^{-23}	1.4388
GHz	3.3356×10^{-2}	1	4.1357×10^{-6}	6.6261×10^{-25}	4.7992×10^{-2}
eV	8.0656×10^{3}	2.41800×10^{5}	1	1.6022×10^{-19}	11605
J	5.0341×10^{22}	1.5092×10^{24}	6.2415×10^{18}	1	7.2430×10^{22}
K	6.9504×10^{-1}	2.0837	8.6173×10^{-5}	1.3807×10^{-23}	1

[1] Technically this conversion between wavelength and wavenumber is only true in vacuum. In other media (for example laboratory air) the refractive index, n_m, of the medium must also be taken into account such that $\widetilde{\nu} = 1/(\lambda_m n_m)$.

T is the temperature in Kelvin, K. At a temperature of 298 K (25 °C) this corresponds to an average relative energy of ca. 200 cm^{-1} and therefore the majority (over 98%) of molecules within a sample are in the ground vibrational state.

2.2 Vibrations of Large Biological Molecules

The vibrations of large polyatomic biomolecules can also be considered, to a first approximation, as harmonic oscillations with a more complex vibrational coordinate based on the motion of individual atoms within the molecule. Not all atomic motions within polyatomic molecules correspond to vibrations, however. Each atom within a molecule has three degrees of freedom, corresponding to displacement along imaginary *x*-, *y*- and *z*-axes. The total number of degrees of freedom of a molecule is therefore $3N$, where N is the number of atoms in the molecule (another way of imagining this is that a total of $3N$ coordinates are needed to define the position of all N atoms in a molecule). The centre of mass of the molecule is defined by three coordinates, and motion of the centre of mass is equivalent to translation: a molecule has three translational degrees of freedom. Additional degrees of freedom because of molecular rotations arise from displacements of atoms about the *x*-, *y*- and *z*-axes. Linear molecules have two rotational degrees of freedom and nonlinear molecules have three rotational degrees of freedom. Of the total $3N$ degrees of freedom, $3N-6$ correspond to molecular vibrations of a nonlinear molecule ($3N-5$ for a linear molecule). These vibrations are known as normal modes, and each atom involved in a particular normal mode oscillates in phase and with the same frequency.

Even relatively small biological cofactors have large numbers of normal modes. For example, the common biological cofactor haem B contains 75 atoms and therefore has 219 normal modes. Myoglobin (a fairly small, ~17 kDa, protein whose functional component is haem B, Fig. 1a) contains approximately 2,940 atoms and has over 8,800 normal modes. It is therefore unrealistic to expect to resolve all the vibrational bands arising from the majority of biological molecules. Fortunately the analysis of vibrational spectra is greatly simplified by the fact that different functional groups or structural motifs vibrate within distinct energy ranges. If groups of atoms within molecules have very different masses, or are connected by bonds of very different stiffness, they can effectively be considered as separate oscillators. For example, because of the small mass of the hydrogen atom, vibrations of atom–H bonds are generally mechanically separate from the rest of the molecule and tend to appear within well-defined spectral regions. The same is true when ligands or small molecules are bound to transition metals, as is the case with the axial His 93 ligand in myoglobin (Fig. 1a) and when CO binds to the Fe ion in the haem B centre to form carboxymyoglobin (Fig. 1b). Vibrational spectroscopy therefore provides information at the sub-molecular level on the structure of biomolecules. Some examples of vibrations commonly encountered in biological

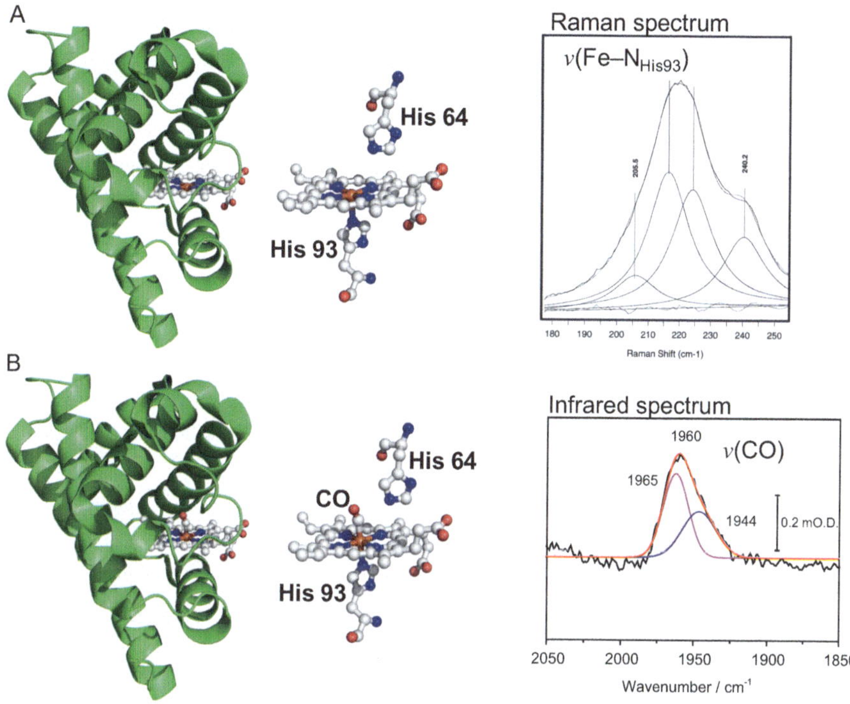

Fig. 1 Ligand binding to haem proteins can be studied using vibrational spectroscopy. (**a**) De-oxy myoglobin (PDB 5D5R), with an expanded view of the region around haem B and a Raman spectrum showing the stretching vibration of the Fe^{II}–N_{His93} bond. (**b**) Carboxymyoglobin (PDB 1DWR), with an expanded view of haem B and an infrared spectrum showing the stretching vibration of the coordinated carbon monoxide ligand. The IR spectrum of carboxymyoglobin has two overlapping CO stretching bands because the coordinated CO has different tilt angles relative to the plane of haem B depending upon the protonation state of His 64: the relative intensities of the two bands vary with solution pH [7]. Raman spectrum adapted with permission from Peterson et al. [8]. Copyright (1998) American Chemical Society

systems are given in Fig. 2a which shows an example of a biological infrared spectrum with peaks assigned to different cell components [9]. Figure 2b shows the protein backbone vibrations which contribute to the amide I, II and III bands. The amide I band is dominated by stretching of the amide carbonyl group, with a small contribution from the N–H bending vibration. The position and shape of the amide I band is very sensitive to the secondary structure of the protein backbone and so is commonly used for analysis of secondary structure [10, 11]. The amide II and III bands are combinations of the N–H bending vibration and the C–N stretching vibration of the amide group; amide II is the out-of-phase combination and amide III arises from the in-phase combination. Analysis of the amide stretching region can be complicated by absorbances of individual amino acid side chains [12]. These absorbances can also be useful, however, as amino acid side chains are often intricately involved in protein function.

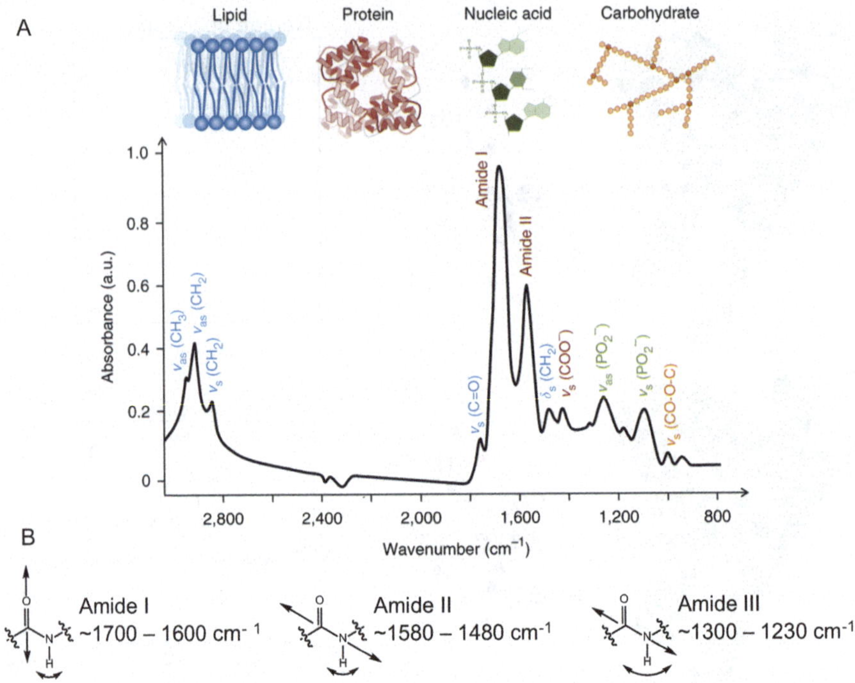

Fig. 2 (**a**) Representative infrared spectrum of human breast carcinoma cells, showing distinct spectral regions in which C–H stretches of lipids, amide bands of proteins, phosphate groups of nucleic acids, and carbohydrates absorb. Reproduced with permission from Baker et al. [9]. Copyright (2014) Nature Publishing Group. (**b**) Vibrational motions of the protein backbone that contribute to the amide I, II and III bands

Conceptually, the mechanical separation of individual oscillators within a molecule can be understood by considering a series of masses linked by springs, such as that illustrated for a 'linear triatomic molecule' in Fig. 3. If all the masses are equal, and the force constants of both springs are identical (Fig. 3a), the nature of the vibration depends on the initial state of the system. If the central mass is fixed at its equilibrium position and both springs are extended equally, then releasing all three masses simultaneously results in symmetric stretching of the 'molecule' and the central 'atom' remains stationary. If the central mass is fixed and only the right-hand spring is extended, then releasing all three masses results in asymmetric stretching of the 'molecule' and the central 'atom' oscillates about its equilibrium position as the right-hand oscillator drives the left-hand oscillator into resonance. Thus the two oscillators are coupled as the behaviour of one is determined by the behaviour of the other. If the system is made asymmetric by introducing an infinite mass (Fig. 3b) then the resonant frequencies of both oscillators are very different and releasing the non-infinite mass causes both springs to distort, but the infinite mass remains stationary. The other extreme case involves a spring with an infinite

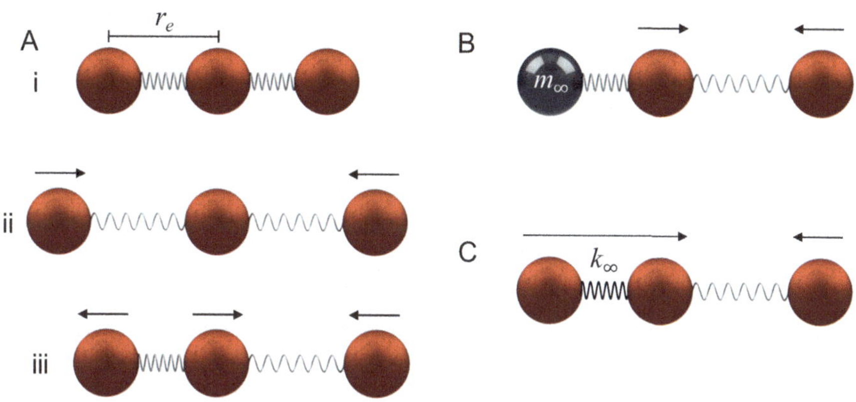

Fig. 3 The coupling of vibrational motions is dependent upon the relative masses of atoms. (**a**) (i) A model symmetric 'linear triatomic molecule', with equal masses connected by bonds of equal stiffness, contains coupled oscillators with identical resonant frequencies; either symmetric (ii) or asymmetric (iii) stretching vibrations can be induced depending upon the initial state of the system. If the 'molecule' is made asymmetric by introduction of an infinite mass (**b**) or infinitely stiff bond (**c**) the resonant frequencies of both oscillators are very different. *Arrows* denote instantaneous direction of motion; *minf* and *kinf* represent an infinite mass or infinitely stiff bond, respectively

force constant (Fig. 3c), and now releasing the right-hand mass causes all the atoms to move, but the infinitely-stiff 'bond' does not distort.

3 Application of IR and Raman Spectroscopy to Bioelectrode Surfaces

3.1 Challenges

Both direct absorption of infrared light and Raman (inelastic) scattering are weak effects. Spectroscopic sensitivity can present a technical challenge even for studies of proteins in solution because it is difficult to prepare high-concentration (>~1 mM) protein samples that retain their native functionality because of aggregation or irreversible denaturation. The sensitivity requirement becomes even more severe when studying proteins adsorbed on electrode surfaces, as we are now recording spectra from a much smaller effective volume, a 'buried' interface (the electrode/electrolyte interface) rather than a bulk solution. The surface coverage of proteins adsorbed on electrodes is typically quite low. Even a densely packed monolayer of a protein with an average diameter of 5 nm corresponds to a maximum surface coverage of the order of 8 pmol cm^{-2}, and uncertainty in protein orientation and stability once immobilised on the electrode means that the *electroactive* coverage may be somewhat lower (in the absence of an orientation-specific immobilisation strategy). Electroactive coverages of moderately sized proteins are often below 1 pmol cm^{-2} in protein film electrochemistry studies,

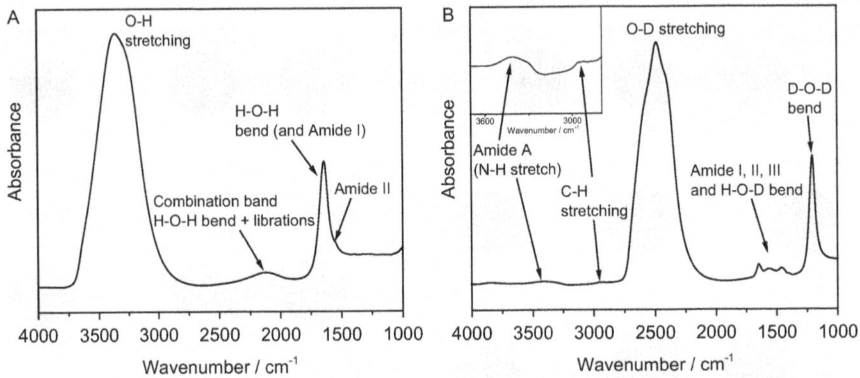

Fig. 4 Water as solvent can cause problems in biological infrared spectroscopy. Isotopic labelling can help to overcome some of these problems, as demonstrated for myoglobin in solution. (**a**) Spectrum of 0.5 mM myoglobin in H_2O, where the amide bands are almost completely obscured by the H–O–H bending vibration. (**b**) Spectrum of 0.5 mM myoglobin in D_2O, showing how the amide bands are now clearly visible. Spectra were recorded in an ATR-IR geometry using a Ge internal reflection element

and therefore high-sensitivity spectroscopic approaches are required when combining protein film electrochemistry with vibrational spectroscopy.

In addition to the absolute sensitivity requirement caused by low electrode surface coverages of proteins, spectroscopic techniques combined with protein film electrochemistry must be able to distinguish between signals arising from small quantities of protein in the presence of relatively large quantities of solvent. In biological vibrational spectroscopy the solvent is necessarily water, H_2O, and this poses a particular problem in infrared spectroscopy because of the intense[2] and broad absorptions of water throughout the mid-infrared [13]. Bands produced by O–H stretching modes appear above 3,000 cm^{-1}, H–O–H bending modes around 1,645 cm^{-1} and a less intense band centred around 2,100 cm^{-1} arising from a combination of the H–O–H bend with low-wavenumber 'libration' modes (restricted rotations: effectively 'rocking' motions arising from the hydrogen bond network of liquid water which absorb below 900 cm^{-1} and interfere with studies in the far-infrared). Figure 4a shows an infrared spectrum of a 0.5 mM solution of myoglobin in water in the 4,000–1,000 cm^{-1} region, in which the amide I band (1,700–1,600 cm^{-1}) is completely obscured by the H–O–H bending vibration, and the amide II band is barely visible as a small shoulder. One way of solving the problem of water absorption in infrared spectra is to use isotopically labelled water, most commonly D_2O, which shifts the solvent bands to lower wavenumbers (lower relative energies) because of the increase in effective mass (from (14) it can

[2] In fact the molar absorptivities (units M^{-1} cm^{-1}) of the water bands are orders of magnitude *lower* than the more intense bands of biological molecules, but the molar concentration of the water solvent is very high.

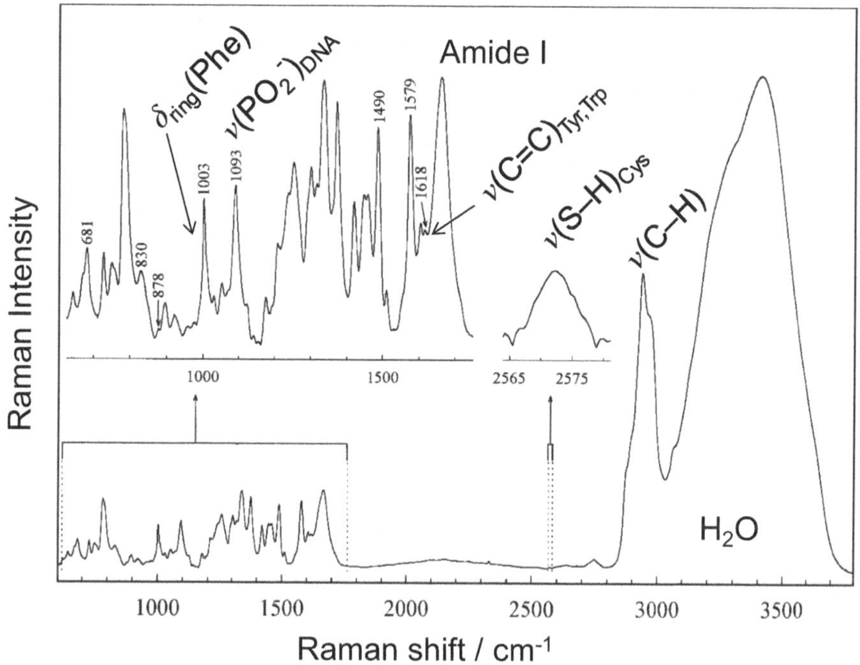

Fig. 5 Raman spectrum of double-stranded DNA bacteriophage P22. Water causes fewer problems in Raman spectroscopy, and much greater detail is visible in the amide regions of the spectrum below 1,700 cm^{-1}. Selected well-defined vibrations of individual amino acids or DNA have been labelled. Reproduced with permission from Thomas [14]. Copyright Annual Review of Biophysics and Biomolecular Structure (1999)

be seen that the wavenumber position is proportional to $\mu^{-1/2}$). This approach can be useful as demonstrated in Fig. 4b which shows another infrared spectrum of a 0.5 mM myoglobin solution, this time in D_2O, where the amide bands are clearly visible. Care has to be taken when exchanging solvent from H_2O to D_2O however, as the amide hydrogen atoms are susceptible to exchange with deuterium which causes a small shift in position of the amide I band (because of a small contribution from N–H bending), and large shifts in amide II and amide III bands (Fig. 2). Not all amide protons are equally susceptible to exchange because of differences in basicity and solvent accessibility, and analysis of spectra can be complicated in samples with partially-exchanged protons, or where proton–deuterium exchange is occurring during the spectroscopic measurement.

Water as solvent is much less of a problem for Raman spectroscopy, as vibrations of the water molecule have weak Raman cross sections. Figure 5 shows the Raman spectrum of the icosahedral double-stranded DNA bacteriophage P22 (a large nucleoprotein complex) in aqueous solution [14]. The spectrum contains clearly resolved bands which can be assigned to individual vibrations, selected examples are indicated in the figure, and there is much less interference from water than in the infrared spectra in Fig. 4.

There are other challenges associated with Raman spectroscopy because of the requirement for a monochromatic (laser) light source. Common laser sources used for Raman spectroscopy are in the visible or near infrared, with typical wavelengths in the region of 488 nm (blue), 532 nm (green), 633 nm (red), 785 nm or 1,064 nm (near-IR). The use of laser light sources can lead to unwanted photophysical and photochemical effects, or even damage to a sample, particularly when using shorter wavelength (higher energy) lasers. If the energy of the laser light matches the energy of an electronic transition, direct absorption of this light excites the molecule from its ground electronic state to a higher electronic state. Because the energy required to promote a molecule to an electronically excited state is greater than the energy required to excite molecular vibrations, it is likely the molecule is also vibrationally excited in the electronically excited state. Overall, the molecule is said to be in an excited vibronic state, from which it can undergo further processes which may or may not involve (re-)emission of radiation and can interfere with the measurement of Raman spectra, or cause fundamental changes to the molecule.

Fluorescence is the emission of light as an electronically excited molecule falls back to the electronic ground state. Resonance fluorescence occurs when light of the same energy as the incident light is re-emitted, and no energy is transferred to the surroundings. More commonly, 'radiationless' transitions occur which transfer thermal energy to the surroundings at the expense of vibrational energy in the molecule. Fluorescence can then occur from the lowest vibrational level of the electronically excited state emitting light of a lower energy. Fluorescence can cause significant problems in the collection of Raman spectra as the energy of the emitted light is shifted towards the energy range of Stokes Raman scattering. Unfortunately, biology is full of fluorescent molecules such as flavin cofactors, nicotinamide adenine dinucleotide cofactors and chlorophylls which are either present in the molecule of interest or can be present as impurities. Even small contributions from background fluorescence can obscure Raman signals so methods must be used that either minimise the magnitude of fluorescence, allow fluorescence contributions to be subtracted from the measured spectra [15] or prevent the fluorescent light being detected [16]. One method for removing the effect of fluorescence on Raman spectra is fluorescence quenching, where an additional component is introduced to the system to accept energy from the excited state of the probe molecule. Metals have a high density of states, and so metal electrode surfaces are quite effective at fluorescence quenching because of good overlap of energy levels with the excited state of the molecule of interest [17].

Molecules in excited electronic states can be susceptible to further chemical reaction. Photochemically-induced isomerisation, proton transfer or electron transfer can occur either as a result of the native function of the biomolecule (for example, photoreceptors such as channelrhodopsins or cryptochromes) or as a side reaction. In biophotoelectrochemistry these properties are exploited or studied (see, for example, Sect. 4.4), but in many other cases it is desirable to avoid these processes as they cause chemical changes to the molecule being studied and obscure spectroscopic details of the state of interest. To minimise the effects of photochemical reactions on Raman spectra their rate must be minimised by choosing a suitable laser power and wavelength without compromising the quality (signal-to-noise ratio) of the spectra. An alternative approach is to use experimental cells in which the sample volume being probed is constantly replaced with fresh, unreacted material, thus allowing higher laser powers to be used. A sample residence time lower than ~40 μs is usually sufficient to avoid interference from photoinduced processes [1]. The use of rotating or flow cells also helps to avoid excessive laser-induced heating of the sample which occurs because of thermal transfer of absorbed radiation to the molecule via radiationless processes. Such heating is a problem even with samples that are photo-inactive and do not fluoresce, and degradation of the sample caused by localised heating can be a major concern in biomolecular Raman spectroscopy.

The measurement of electrochemical signals (current, potential) is a crucial part of any combined direct electrochemical-spectroscopic experiment and any experimental cell design involves a compromise between optimal collection of spectroscopic and electrochemical data. Electrochemically, it is desirable to use a three-electrode electrochemical cell in which uncompensated resistance is minimised by placing the reference electrode in close proximity to the working electrode (uncompensated resistance arises because of the finite size of a reference electrode placed at a finite distance from the working electrode) [18]. The effect of uncompensated resistance is to cause an error in potential control whenever a Faradaic process takes place in the electrochemical cell (i.e. whenever a current is flowing at the working electrode). The magnitude of the error in potential is proportional to the measured current (because $V = iR$) and so can become significant when high catalytic currents are measured from samples of highly active redox enzymes, particularly at large electrodes such as those often used when coupling vibrational spectroscopy and bioelectrochemistry. As a result, it is often desirable to use miniaturised reference electrodes, such as Ag/AgCl or saturated calomel electrodes. In cases where it is not possible to miniaturise the electrode enough to fit into the spectroscopic cell, a silver wire pseudo-reference can be used, but the use of a pseudo-reference is at the expense of increased potential drift during an experiment. The counter electrode is typically platinum, and the surface area of the counter electrode must be large enough to ensure that the rate of any Faradaic process of interest at the working electrode is not limited by the capability of the counter electrode to balance the current passed. This means that a second Faradaic process occurs at the counter electrode, which is generally ignored, although oxidation of water at a platinum counter electrode to compensate a reductive reaction at the working electrode results in the production of trace O_2. In cases where the presence of trace O_2 is undesirable, for example when studying highly oxygen-sensitive species, the counter electrode must be separated from the main electrochemical cell in a separate compartment, using a frit or membrane. The position of the counter electrode is also important and, to ensure homogeneous current flow, it should ideally be equidistant from all points on the surface of the working electrode.

3.2 Sampling Geometries Compatible with Direct Electrochemistry

Despite the inherent challenges imposed by spectral sensitivity, sample handling and electrochemical cell design, IR and Raman spectroscopies are very convenient tools for the study of electrode-immobilised biomolecules. This convenience is, at least in part, because of the compatibility of vibrational spectroscopies with a range of sampling geometries, most of which can be modified to accommodate electrochemical measurements (Fig. 6).

A transmission geometry (Fig. 6a) is perhaps the most familiar arrangement. There are a number of examples of the use of IR spectroscopy in a transmission geometry to study redox proteins in solution [19, 20], using variations on an electrochemical sample cell introduced by Mäntele and co-workers [21]. The application of transmission IR spectroscopy to electrode immobilised proteins is made difficult by the requirement for electrodes transparent to IR light, compatible with biomolecules over a wide potential range, and the limitation of using a thin-layer electrochemical cell. The requirement for a thin-layer cell is imposed by the strong background absorbance of water which places a practical upper limit on the pathlength of a transmission cell to a few tens of micrometres. Thin-layer electrochemical cells typically suffer from large uncompensated resistances and therefore an offset in cell potential, together with difficulties in mass transport (reactant supply and product removal) which would be a problem with very active redox enzymes. A range of optically transparent electrode materials, including boron doped diamond [22], indium tin oxide [23], TiO_2 [24] and SnO_2 [25, 26], have been successfully applied to the study of haem proteins using UV-visible and MCD spectroscopies. Transmission Raman spectroscopy was re-introduced as a viable experimental approach by Matousek and Parker in 2006, but is mainly used for the analysis of pharmaceutical tablets and biological tissue samples [27, 28].

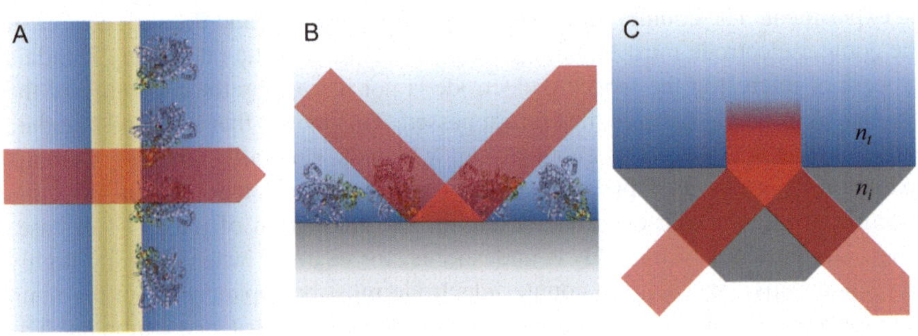

Fig. 6 Sampling geometries available for coupling vibrational spectroscopy and electrochemistry. (**a**) Transmission through a transparent working electrode. (**b**) External reflection from a reflective electrode. (**c**) Attenuated total reflectance with a working electrode deposited on the surface of an internal reflection element

Reflection-based approaches are most commonly used to couple IR spectroscopy with direct electrochemistry, using either IR reflection-absorption spectroscopy (IRRAS,[3] Fig. 6b) or attenuated total reflectance IR (ATR-IR, Fig. 6c) spectroscopy. In IRRAS the incident IR light travels through solvent, the less optically dense (lower refractive index) medium, before being reflected from the electrode surface, the more optically dense (higher refractive index) medium. The IRRAS geometry is suited to cases where the working electrode can be polished to a mirror-like finish, most commonly using metal electrodes, to minimise reflection losses which would limit spectral quality (signal-to-noise ratio). Spectroscopically the application of IRRAS is challenging because of the small volume of material sampled at the electrode surface relative to the bulk solution which the IR radiation must pass through. The presence of highly absorbing reactants, products or buffer components can complicate the interpretation of spectra. Because both the incident and reflected IR light passes through water, bioelectrochemical applications of IRRAS suffer from similar electrochemical difficulties to transmission IR spectroscopy in terms of uncompensated resistance and mass transport limitations of a thin-layer electrochemical cell. Despite these drawbacks, IRRAS, and in particular an intrinsically surface sensitive variation PM-IRRAS (polarisation modulation IRRAS), has been used to study the effect of electrode potential on protein structure, mainly in the amide I and II regions [29, 30].

In ATR-IR spectroscopy (Fig. 6c) the incident IR light travels through an optical component known as an internal reflection element (IRE) and is totally internally reflected at the IRE/sample interface, where the IRE has a higher refractive index than the sample. Total internal reflection occurs at angles of incidence greater than the critical angle, θ_c, which can be determined from Snell's law (15) as the angle of incidence (θ_i) at which the angle of transmittance (θ_t) becomes 90° (16) [6]. In these equations, n_i and n_t are the refractive indices of the IRE and the sample, respectively, as defined in Fig. 6c.

$$n_i \sin\theta_i = n_t \sin\theta_t, \tag{15}$$

$$\theta_c = \arcsin n_t/n_i, \tag{16}$$

Despite the term total internal reflection, the electromagnetic wave of the reflected light does extend into the sample, known as an evanescent wave. If the wavelength of any of the incident light corresponds to an absorption band of the sample, then some energy is transferred to the sample and the intensity of the reflected light is decreased (the reflectance of the surface is less than 1). The resulting spectrum is effectively an absorption spectrum of the sample and the reflected light is said to have been attenuated. The evanescent wave occurs as a result of a boundary condition which requires the electric field of the incident light perpendicular to the plane of incidence to be continuous across the IRE/sample interface [31]. The

[3] IRRAS is also variously referred to as reflection-absorption IR spectroscopy (RAIRS) or abbreviated to IRAS in the literature.

electric field, E_z, of the evanescent wave decays exponentially with distance, z, from the IRE surface (17) with a characteristic decay length known as the penetration depth, d_p (18).

$$E_z \propto E_0 e^{-z/d_p}, \tag{17}$$

$$d_p = \frac{\lambda}{2\pi n_t \sqrt{\left(\frac{n_i}{n_t}\right)^2 \sin^2\theta_i - 1}}. \tag{18}$$

The penetration depth defines the value of z at which the electric field has decayed to $1/e$ of its (maximum) value at the IRE surface. It can be seen from (18) that the penetration depth is dependent on both the angle of incidence and the ratio of refractive indices of the IRE and sample. For a given sample and angle of incidence (i.e. at constant n_t and θ_i) the penetration depth decreases with increasing refractive index of the IRE, n_i. For a given IRE material (i.e. constant n_i and n_t) the penetration depth decreases with increasing angle of incidence, θ_i. A reasonable estimate for the refractive index of a biological sample in aqueous solution is 1.4, and a range of values of critical angle and penetration depth for a theoretical non-absorbing sample on various common IRE materials are given in Table 2, calculated at a wavelength of 10 μm (a relative energy of 1,000 cm^{-1}) and angle of incidence of 60°. Close to the critical angle, the penetration depth changes sharply with incidence angle. Practically it is favourable to avoid this arrangement by recording spectra at angles of incidence well above the critical angle, especially when trying to carry out quantitative measurements. When considering which IRE material to use, and taking the refractive index into account, it is also important to consider chemical compatibility. For example, ZnS, ZnSe and Ge are commonly used IRE materials, but they are not as robust as either Si or diamond, are not resistant to cleaning with either strong acids or bases and are quite brittle. Diamond, although it is transparent over a wide range of the mid- and far-IR, has an intense IR-active

Table 2 Physical properties of common internal reflection element materials, together with representative penetration depth in water (n_t ~ 1.33, ignoring absorbance) at a wavelength of 10 μm (~1,000 cm^{-1})

	Refractive index, n_i	Critical angle, θ_c/°	Penetration depth, d_p, at $\theta_i = 60°$/μm	Approximate useful range/cm^{-1}
Diamond	2.4	33.65	0.996	30,000–<10[a]
Ge	4.0	19.42	0.498	5,500–500
Si	3.4	23.03	0.606	8,900–600[b]
ZnS	2.2	37.20	1.167	17,000–800
ZnSe	2.4	33.65	0.996	15,000–600

[a]Diamond has a strong lattice vibration between ~1,800 and 2,100 cm^{-1} and is effectively opaque in this region for even moderate pathlengths

[b]Depending upon the crystallisation method and pathlength, the lower wavenumber limit for Si could be as high as ~1,500 cm^{-1}. Si also has a transparent window in the far-IR, between ~300 and 100 cm^{-1}

lattice vibration between ~1,800 and 2,100 cm^{-1}. Diamond is therefore not generally a suitable IRE material for studying the binding of small molecule ligands such as CO, CN^- or NO to metalloproteins, as these diatomic ligands have absorbances in this wavenumber region. Silicon is also transparent over much of the mid-IR but, depending upon the method used to crystallise the raw material, Si IREs can have limited transparency below about 1,500 cm^{-1}.

To gain quantitative information from ATR-IR spectra it is necessary to relate the quantity actually measured, the reflectance, $R = I/I_0$, to the absorbance measured in the transmission geometry [31]. Because IR spectroscopy is an absorption spectroscopy it obeys the Beer–Lambert law (19), at least for relatively weak absorbances such as those encountered in biological samples. This means that the absorbance, A, is proportional to the pathlength, D, molar extinction coefficient, ε, and sample concentration, c:

$$A = \log_{10}\frac{I_0}{I} = D\varepsilon c. \tag{19}$$

An equivalent expression can be written for ATR-IR spectroscopy,

$$A = -\log_{10}R = d_{\text{eff}}\varepsilon c, \tag{20}$$

where d_{eff} is the effective depth, which represents the pathlength required to record the same absorbance in a transmission measurement. In the thick film limit, where the thickness of the sample on the IRE is much larger than the penetration depth, the sample can be considered to be homogeneous over the whole evanescent wave. In this case, the effective depth is proportional to the penetration depth, as described by Harrick [32]. The reason why d_p and d_{eff} are not the same arises from the fact that IR absorption is a linear spectroscopy (as is Raman scattering), which means that the signal intensity is proportional to the electric field strength squared. One consequence of this is that the contribution to the measured spectrum is not equal at all points of the evanescent wave and it is really the integral of the electric field strength squared as a function of distance from the IRE surface that determines the measured reflectance. (The square of the electric field decays twice as quickly as the electric field itself, and so ATR-IR spectroscopy actually probes a sample depth with a characteristic length of $d_p/2$.) Fortunately in practice the penetration depth and effective depth are often very similar, particularly for weakly absorbing samples on top of high refractive index IREs.

The use of an ATR-IR geometry has distinct advantages over transmission and IRRAS when coupling to an electrochemical measurement. Because the IR beam does not have to pass through the sample, the use of thin-layer electrochemical cells can be avoided if the working electrode is located at the IRE surface. Introduction of reactants and inhibitors, or changing other solution conditions such as pH, ionic strength or buffer composition, is facile if a protein is immobilised in a film at the IRE surface. Mass transport limitations can be overcome through the use of flow cells, or by stirring the bulk electrolyte. Even highly absorbing samples can be

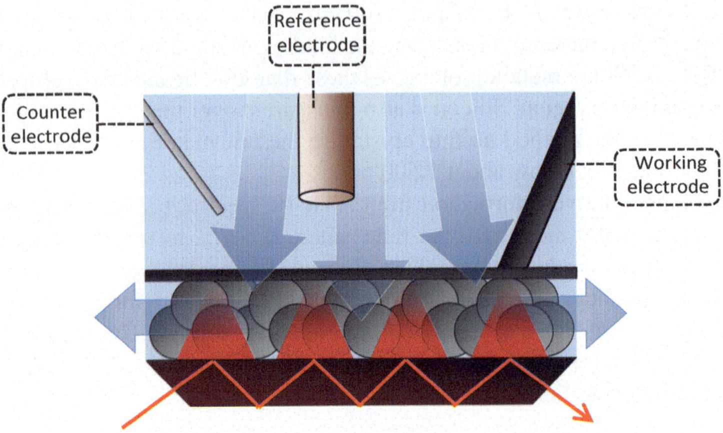

Fig. 7 Schematic cell design used for protein film IR electrochemistry (PFIRE) experiments. Protein is adsorbed onto carbon particles which are then deposited onto an Si internal reflection element. Covering the deposited particles with carbon paper provides good electrical connection to the working electrode. Solution flow, indicated by *arrows*, provides fresh substrate and prevents build-up of product in the working electrode film. Reproduced with permission from Hidalgo et al. [33]. Copyright 2015 The Authors. Published by Wiley-VCH Verlag GmbH & Co. KGaA, Weinheim

measured using ATR-IR spectroscopy in conjunction with a high refractive index IRE and so some of the optical constraints on electrode materials are also removed. Despite carbon being opaque, Vincent and co-workers have developed a method known as protein film infrared electrochemistry (PFIRE, Fig. 7) in which enzymes are adsorbed onto high surface area carbon particle working electrodes deposited onto the surface of an Si IRE [33]. Carbon working electrodes, commonly used in protein film electrochemistry, are chemically inert over the biologically accessible potential range and have high overpotentials for the activation of small molecules such as H_2 and O_2. The PFIRE method makes use of ATR-IR spectroscopy approaching the thick film limit as the thickness of the particle electrode is of the order of several micrometres and the penetration depth is less than 0.5 μm at 2,000 cm^{-1}. High spectroscopic sensitivity is achieved by the large surface area carbon (up to 1,300 m^2 g^{-1}) used to fabricate the working electrode; effective protein concentrations approaching 1 mM can be achieved within the penetration depth of the evanescent wave, even with surface coverages as low as 1 pmol cm^{-2}. Spectral sensitivity is increased further by the use of a multiple reflection IRE, which increases the effective depth (and therefore sample absorbance in accordance with (20)) by a factor N, where N is the number of reflections.

3.3 Surface Enhanced Vibrational Spectroscopies

The magnitude of Raman scattering and IR absorbance can be enhanced dramatically when a monolayer protein film is adsorbed on a rough metal surface or metal nanostructure [34, 35]. This effect is exploited by both surface enhanced Raman spectroscopy (SERS) and surface enhanced infrared absorption (SEIRA) spectroscopy, where the metal surface can be used as the working electrode in an electrochemical cell. The origin of the surface enhancement is related to driven oscillation of delocalised conduction electrons in a metal nanoparticle upon illumination by an external light source [36]. In classical terms, the oscillating electric field of the light source applies a periodic force to the conduction electrons which in turn induces periodic charge separation, known as localised surface plasmon resonance, in the nanoparticle (Fig. 8). In effect, an oscillating dipole moment, μ_{ind}, has been induced, which acts as a nanoantenna emitting radiation of the same frequency as the external light source. The net result is resonant elastic light scattering from the nanoparticle, and an enhanced local electric field close to the nanoparticle surface. The intensities of Raman scattering and infrared absorption are proportional to the square of the electric field, and so we have described qualitatively the electromagnetic enhancement mechanism of SERS and SEIRA. Electromagnetic enhancement is dependent on the intensity and wavelength of the incident light and the polarisability of the metal nanostructure, factors that affect whether the metal has a plasmon resonance in a useful spectral region. For biological spectroscopy nanoscale silver and gold can be used, as these materials have plasmon resonances in the visible to near infrared.

Although the electromagnetic enhancement mechanism can be used for a conceptual understanding of both SERS and SEIRA, the fundamental characteristics of the two techniques are different. The intensity enhancement in SEIRA depends only

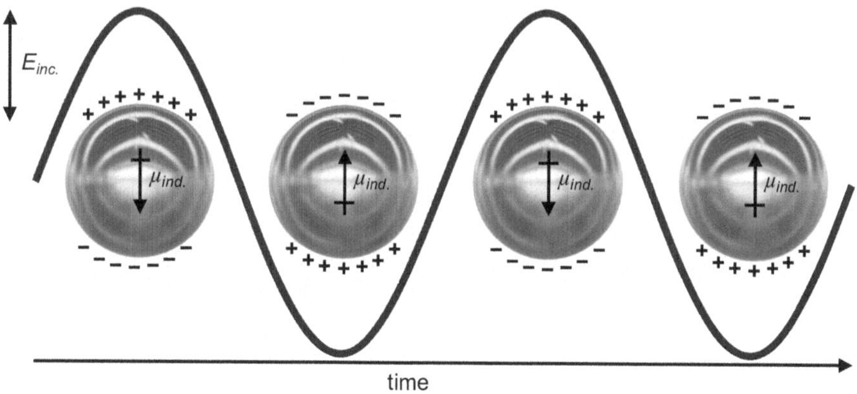

Fig. 8 Localised surface plasmon resonance in a nanoparticle induced by illumination with an external light source. An oscillating dipole, μ_{ind}, is induced in the nanoparticle which results in emission of radiation and enhancement of the electric field close to the nanoparticle surface

on the increased local electric field caused by the illuminating light because IR is an absorption spectroscopy and so scales approximately with the square of the incident light intensity only. The enhancement is much greater for SERS, however, as the Raman scattered light from the probe molecule can also excite local surface plasmon resonance in the metal support (in general the energy difference between the laser light used for excitation and the Raman scattered light is small and so both are likely to be in resonance with plasmons in the metal). In the case of SERS, therefore, the intensity enhancement scales approximately with the fourth power of the incident light intensity, E^4, and even small local electric field enhancements lead to significant increases in SERS intensity. For example, an increase of $E_{\mathrm{local}}/E_{\mathrm{incident}} = 100$ leads to a SERS enhancement of one hundred million.

Practically, rough silver or gold surfaces needed for SERS and SEIRA measurements are produced by chemical deposition, evaporation or electrochemical roughening of electrodes through successive oxidation and reduction cycles [37]. Electrochemical cells for SERS applications tend to employ an external reflection-like geometry, where Raman scattered light is collected at either 90° or 180° (backscattered light) [38]. Unwanted photoinduced processes are also enhanced by the high local electric field at the electrode surface and so SERS cells employ eccentrically rotating ring or disc electrodes to ensure continuous and rapid motion of the sample through the focus of the incident laser. Electrochemical cells for SEIRA tend to use an ATR configuration with 20–200 nm thick gold films deposited onto single-reflection Si IREs, and hence SEIRA is often referred to as ATR-SEIRA [37, 39]. An alternative, more complex, description of the ATR-SEIRA effect considers how the refractive index of a thin gold film is modified by adsorbed protein, which leads to changes in reflectance around vibrational modes of the adsorbed protein [34].

Direct interaction between bare metal surfaces and proteins can lead to denaturation through strong interactions of thiol groups in the protein with the metal surface. Metal electrodes are therefore generally protected with self-assembled alkanethiol monolayers to prevent direct contact with biomolecules (see Sect. 2.1 of [40]). Proton reduction and desorption of the thiol monolayer imposes a lower limit on the accessible redox potentials of approximately −400 mV vs the standard hydrogen electrode for gold and −300 mV for silver and so it is difficult to study the low-potential domain of biological redox processes. Use of self-assembled monolayers (SAMs) also has implications for spectroscopic measurement, as the strength of electromagnetic enhancement decreases with distance from the electrode surface. The enhancement factor scales as $1/z^3$, where z is the distance from the electrode surface, in line with the distance-dependence of the electric field strength of dipolar radiation [36]. An expression for the decay of SEIRA enhancement, $F_{\mathrm{SEIRA}(z)}$, can then be written in terms of the SEIRA enhancement at the surface, $F_{\mathrm{SEIRA},0}$, and the average roughness of the surface, r:

$$F_{\mathrm{SEIRA}(z)} = F_{\mathrm{SEIRA},0}\left(\frac{r}{r+z}\right)^{6}. \tag{21}$$

An equivalent expression can also be written for the decay in SERS enhancement, but in this case the decay is much more rapid because of the E^4 dependence of the SERS effect:

$$F_{\mathrm{SERS}(z)} = F_{\mathrm{SERS},0}\left(\frac{r}{r+z}\right)^{12}. \tag{22}$$

Figure 9 shows the calculated decay in signal enhancement from a surface with an average roughness of 20 nm for both SERS (Fig. 9a, green solid line) and SEIRA (Fig. 9a, red dashed line) alongside two representative proteins (a two-subunit NiFe hydrogenase, red, and cytochrome *c*, green, a small electron transfer protein) immobilised on a 'SAM' (grey) of 1 nm thickness. The signal enhancement decays rapidly over a length-scale similar to the size of protein molecules, and so not all parts of the protein contribute equally to the spectra. Applications of SERS or SEIRA to large, multi-subunit proteins can therefore be challenging [41]. Even in

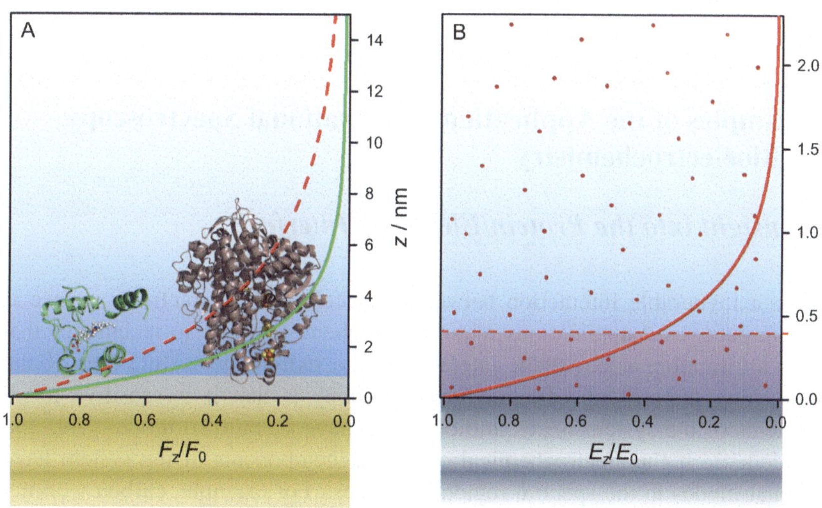

Fig. 9 The distance dependence of surface enhancement for SERS (**a**, *green solid line*) and SEIRA (**a**, *red dashed line*) are on the same order as the size of biomolecules, whereas the evanescent wave in ATR-IR spectroscopy (**b**) extends into bulk solution. Protein structures are shown to scale either adsorbed on a 1 nm self-assembled monolayer (**a**, *grey*) or in solution (**b**) and are *Escherichia coli* nickel-iron hydrogenase 1 (*red*, PDB code 1WUH) and equine cytochrome *c* (*green*, PDB code 3NBS). The SERS and SEIRA decay factors, F_z/F_0, are the fractional enhancements at distance, *z*, relative to the maximum enhancement at the metal surface ($z = 0$) and were calculated according to (21) and (22) with a surface roughness of 20 nm. The electric field decay in (**b**) was calculated from (18) for a wavelength of 5 µm (2,000 cm^{-1}), an incidence angle of 45° on an Si IRE ($n_i = 3.4$) and pure water solvent ($n_t \sim 1.33$)

the case of the small cytochrome *c* molecule, the intensity of amide features in SEIRA spectra is found to depend on the protein orientation because of the spectra being weighted towards features close to the electrode surface. Protein orientation has an additional effect, as the magnitude of surface enhancement is also dependent on the orientation of the change in dipole moment of a vibration relative to the surface [34, 38]. Vibrations whose change in dipole moment lie almost parallel to the surface are more weakly enhanced, and so by comparing the relative intensities of bands with perpendicular changes in dipole moment it is possible to probe protein orientation. Hildebrandt and co-workers exploited this property to monitor changes in orientation of cytochrome *c* electrostatically adsorbed onto self-assembled monolayers in response to potential steps using vibrational modes of the planar porphyrin chromophore [42]. Figure 9b shows, for comparison, the penetration depth probed by ATR-IR spectroscopy relative to the same NiFe hydrogenase in solution. A major advantage of SERS and SEIRA is that they are near-field optical techniques; they are very sensitive to changes in monolayers at the electrode surface but are relatively insensitive to changes in the bulk solvent. In contrast, ATR-IR measurements are extremely sensitive to the bulk solvent because of the much larger penetration depth. SERS and SEIRA measurements are therefore advantageous in cases where it is desirable to study spectral changes in the amide or O–H stretching regions which could otherwise be obscured by bulk water absorbances.

4 Examples of the Application of Vibrational Spectroscopy to Bioelectrochemistry

4.1 Insight into the Protein/Electrode Interface

Forming a favourable interaction between a protein and an electrode surface is critical to successful applications of bioelectrochemistry, and so techniques that provide information about these interactions are extremely useful. One of the distinguishing features of vibrational spectroscopy is that it is sensitive to all components of the system being studied which have vibrational modes. This can be problematic if the electrochemical or optical components of the system have vibrational modes in the spectral region of interest. For example, carbon working electrodes contain oxygenated surface functionalities that absorb close to the amide I and amide II regions, and optical materials such as silicon or diamond have characteristic vibrations that cause interference in certain ranges of the mid-IR. Buffer components such as phosphates or sulphates have absorbances in the same region as some biological cofactors or amino acid side chains. The ability to 'see everything' can, however, also be an advantage, as in some situations it allows characterisation of every step in bioelectrode formation in situ.

Affinity chromatography is widely used to purify protein samples [43], and the same approach can be used to form stable monolayers of proteins on electrode surfaces. The use of polyhistidine-tagged proteins and an electrode surface modified with nitrile-triacetic acid (NTA) terminated SAMs is one such example, and the whole process from SAM formation to protein immobilisation and spectroelectrochemical characterisation can be followed spectroscopically (see also Sect. 2.2 of [40]). Figure 10 shows ATR-SEIRA spectra recorded during the

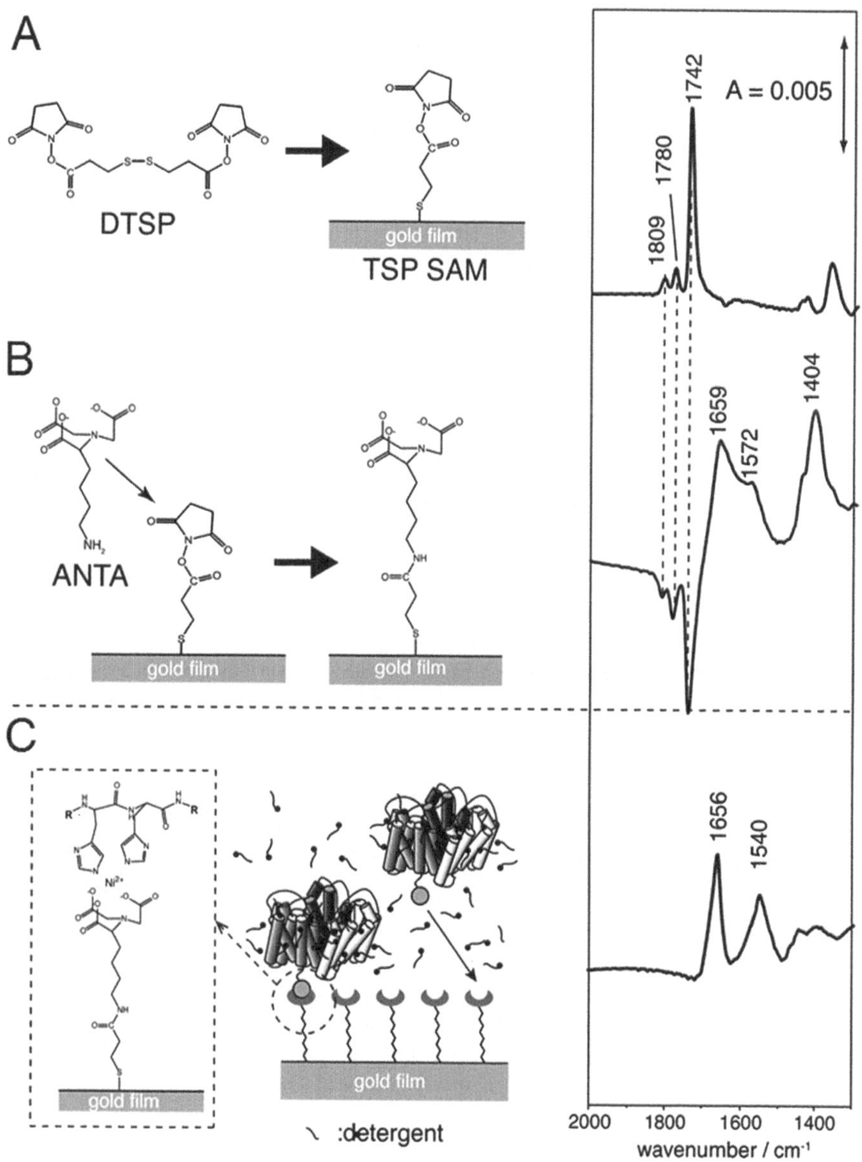

Fig. 10 Surface modification with a NTA-terminated self-assembled monolayer (**a**, **b**) and adsorption of cytochrome *c* oxidase (**c**) monitored by ATR-SEIRA spectroscopy. Reprinted from Ataka et al. [34] with permission from Elsevier

formation of a NTA-terminated SAM and immobilisation of cytochrome *c* oxidase [34, 44]. Reaction of a disulphide-containing molecule, dithiobis(succinimidyl propionate) (DTSP), with a gold surface (Fig. 10a) results in formation of a thiosuccinimidyl propionate (TSP) adduct covalently bound at the surface as evidenced by a strong band at 1,742 cm^{-1}, characteristic of an ester carbonyl stretch, in the SEIRA spectrum. Reaction of the TSP monolayer with a chelator, amino-nitrilotriacetic acid (ANTA), leads to loss of absorbances because of the ester group (negative peaks) and gain of features characteristic of an amide linkage and carboxylate groups (positive peaks), showing incorporation of ANTA into the monolayer (Fig. 10b). The SEIRA spectrum in Fig. 10b demonstrates the usefulness of difference spectra in vibrational spectroscopy; by subtracting a spectrum recorded before a reaction from a spectrum recorded after a reaction it is possible to isolate only the spectral features that have changed as a result of the reaction. Addition of Ni^{2+} and His-tagged cytochrome *c* oxidase results in a stable monolayer of immobilised protein (Fig. 10c) with clear amide I and amide II bands which are resistant to rinsing unadsorbed protein from solution. It is then possible to reconstitute the adsorbed cytochrome *c* oxidase into a lipid membrane. Residual curvature to the high wavenumber side of the amide I band in Fig. 10c is probably because of the thickness of the particular gold film deposited at the beginning of the experiment [45]. Addition of cytochrome *c*, the native redox partner of cytochrome *c* oxidase, to solution above the ATR prism led to significant oxygen reduction at applied potentials below +260 mV vs the normal hydrogen electrode as binding of cytochrome *c* resulted in electron transfer to the cytochrome *c* oxidase catalytic centre, followed by oxygen reduction [46]. These results indicate that immobilised cytochrome *c* oxidase retained native functionality, an essential criterion for successful spectroelectrochemical studies of immobilised proteins. Furthermore, potential-induced difference spectra in the amide region showed that conformational changes in cytochrome *c* in solution and bound to the cytochrome *c* oxidase surface were very similar.

Care must be taken when designing SAMs for protein adsorption, as proteins adsorb to surfaces via a variety of weak and non-specific interactions. Hydrophobic interactions, hydrogen bonding interactions and electrostatic interactions suffer from sensitivity to changing solution conditions, and even the His-tag approach may lead to unspecific binding if the protein has alternative surface-exposed histidine residues. Therefore there is no guarantee that proteins retain their native conformation on immobilisation. Sezer et al. have demonstrated how the structure and redox properties of haem centres can be sensitive to SAM structure and immobilisation [47]. Surface enhanced resonance Raman spectroscopy (SERRS) was used to investigate the role of a dihaem cytochrome subunit in the electron transfer pathway of *Ralstonia eutropha* membrane-bound hydrogenase. Partial conversion of native six-coordinate low-spin haem to a five-coordinate high-spin state occurred when the protein was immobilised on carboxylic acid or amine terminated SAMs at low ionic strength. Increasing the ionic strength and immobilising on mixed SAMs containing alcohol terminated thiols restored native haem coordination at the expense of protein desorption. Immobilisation of cytochrome *c* oxidase on similar mixed alcohol- and amine-terminated thiol SAMs resulted in a non-native haem species which had a markedly lower redox potential than the native haems in solution. Sezer et al. concluded that electron supply to the catalytic oxygen-reducing centre is disrupted by immobilisation and postulated that this effect may account for the poor performance of immobilised cytochrome *c* oxidase in enzyme fuel cells.

4.2 Effect of Membrane Potential on Protein Structure

Many proteins are either permanently anchored to cell membranes or function through interaction with membranes. Approximately one quarter of the proteins produced by *Escherichia* (*E.*) *coli* are thought to be membrane proteins. Because of the difficulty of membrane protein crystallography, however, only a small percentage of protein structures that have been determined are membrane proteins. Spectroscopic study of functional membrane proteins by incorporation or reconstitution into model bilayer lipid membranes can therefore provide valuable information which is otherwise difficult to obtain.

The existence of a membrane potential is common to all cells, and despite this there are relatively few techniques for investigating the effect of transmembrane potential on the conformation of membrane proteins. The voltage-dependent anion channel is a vital transmembrane protein which governs metabolite transport across the mitochondrial outer membrane and is involved in apoptosis. The activity of the anion channel is dependent on the membrane potential, and Kozuch et al. have used SEIRA to study the effect of membrane potential on conformation of the human voltage-dependent anion channel embedded in electrode-supported bilayer lipid membranes [48]. Incorporation of the anion channel was confirmed by electrochemical impedance spectroscopy, which showed that the resistance of electrode-supported bilayers decreased when the anion channel was present. Figure 11a shows double difference spectra recorded at ± 60 mV vs open circuit potential, where open circuit potential refers to the 'open' conformation of the anion channel. The spectra are presented as double difference spectra to subtract contributions from the lipid membranes, which also have absorbances in the amide region. At applied membrane potentials above open circuit potential there is a positive band in the O–H stretching region as the membrane bends towards the electrode (Fig. 11b), indicating that water molecules are accommodated in the pore of the anion channel. Likewise, the bands at 1,720 and 1,505 cm^{-1} are probably produced by amino acids in the anion channel on the underside of the membrane which change orientation relative to the electrode surface as the potential is stepped over open circuit potential. The bands at 1,560 and 1,529 cm^{-1} are caused by the β-sheet structure of the anion channel, and the fact that the intensity changes of these two bands have opposite sign again suggests an orientation change of the anion channel with applied potential. At applied potentials lower than open circuit potential, the anion channel constricts, either by stretching away from the electrode surface

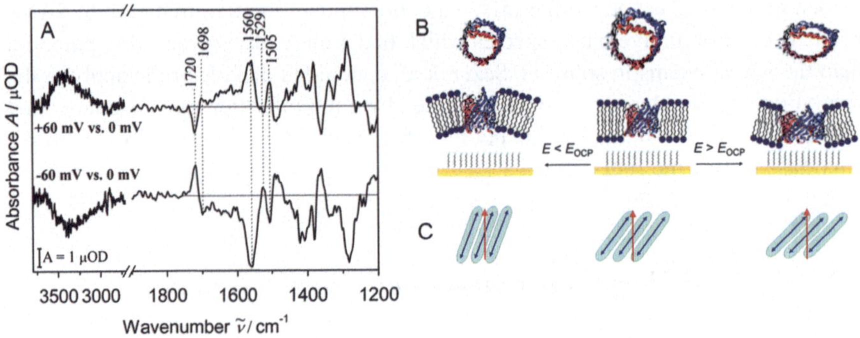

Fig. 11 ATR-SEIRA studies of the effect of membrane potential on the structure of human voltage-dependent anion channel reconstituted into a supported bilayer lipid membrane. (**a**) Double difference spectra showing changes in the O–H stretching and amide regions at potentials above and below open circuit potential. Peaks in the difference spectra can be explained by either stretching and compressing of the anion channel (**b**) or tilting (**c**) of the overall transition dipole moment of the amide bands. Reproduced from [48] with permission from the PCCP owner societies

(Fig. 11b) or by tilting away from the surface (Fig. 11c). The opposite occurs above open circuit potential, and the anion channel distorts as it is compressed or tilts towards the electrode surface. Electrochemical methods alone cannot provide such structural information, as they only show the effect of anion channels in increasing ion transport across the membrane. ATR-SEIRA spectroscopy has also been used to study the KcsA potassium channel [49].

4.3 Redox Processes Occurring at Specific Sites Within Proteins

The transfer of electrons in and between proteins is achieved by redox cofactors – haem centres, iron-sulphur clusters, copper centres or organic molecules such as flavins. In general these redox cofactors have distinct spectroscopic signatures at each accessible redox level, and these signatures can be used to characterise the reduction potential of the cofactor. The nature of the interaction of an outer membrane cytochrome, OmcB, from *Desulfuromonas acetoxidans* with an electrode surface was investigated by Millo and co-workers [50]. *Desulfuromonas acetoxidans* is a metal-respiring bacterium which oxidises organic molecules in water and uses a metal-containing mineral as an electron sink, and OmcB is a multihaem protein which is thought to be involved in this extracellular electron transfer. Figure 12a shows Raman spectra of *Desulfuromonas acetoxidans* in solution (i) and on a silver electrode at potentials of −335 mV (ii) and +45 mV (iii) vs the saturated calomel electrode. The raw spectra (black) have been fitted to account for the presence of both oxidised and reduced forms of six-coordinate

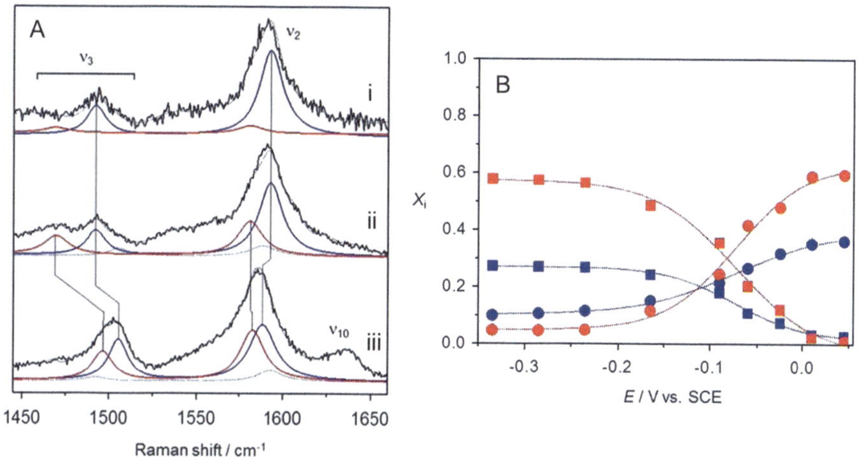

Fig. 12 Redox titration of a multihaem outer membrane cytochrome from *Desulfuromonas acetoxidans*. (**a**) Raman spectra recorded in solution (i) and on a silver electrode at −335 (ii) and +45 (iii) mV vs the saturated calomel electrode (*black*) and spectral fitting into contributions from six-coordinate low-spin (*blue*) and five-coordinate high-spin (*red*) haem cofactors. (**b**) Redox titration of the oxidised (*circles*) and reduced (*squares*) haem cofactors. Reprinted with permission from Alves et al. [50]. Copyright (2015) American Chemical Society

low-spin and five-coordinate high-spin haems following adsorption on the electrode. Repeating this fitting at a range of applied potentials gives a redox titration of all the haems in OmcB, as shown in Fig. 12b. The same approach can be extended to other redox cofactors, and Moe et al. have shown that surface enhanced resonance Raman spectroscopy can be used to study the redox chemistry of Endonuclease III, a small iron-sulphur-containing enzyme [51].

4.4 *Biophotoelectrochemistry*

Many biological processes are sensitive to, or triggered by, visible light illumination. Even in cases where this illumination does not lead directly to a redox process it can lead to structural changes in electrode immobilised proteins, the magnitude of which are sensitive to the electrode potential. Structural changes range from relatively localised proton transfers to photolysis of exogenous ligands, and in some cases can lead to large changes in amide band structure.

ATR-SEIRA spectroscopy has been used by Jiang et al. to study the effect of transmembrane potential on light-triggered structural changes in the integral membrane protein sensory rhodopsin II [52]. Absorption of blue-green light by the chromophore retinal (Fig. 13a (i, ii)) triggers conformational changes in sensory rhodopsin II which ultimately lead to the halobacteria moving away from the light, propelled by a flagella motion. Figure 13b shows light *minus* dark difference spectra

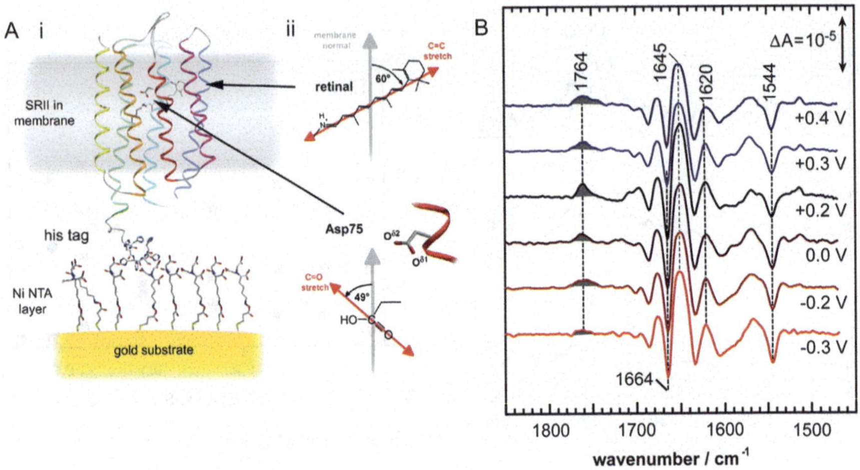

Fig. 13 Photo-induced proton transfer in the transmembrane protein sensory rhodopsin II. (**a**) (i) Reconstitution of the protein in a lipid membrane and (ii) orientations of retinal and an aspartate amino acid which are the proton donor and acceptor, respectively. (**b**) Light *minus* dark difference spectra recorded at different applied membrane potentials. Reproduced with permission from Jiang et al. [52]

recorded at a range of applied electrode potentials. The negative vibrational band at 1,544 cm^{-1} is the C=C stretch of the photo-isomerised retinal Schiff base, which transfers a proton to the carboxylate side chain of a nearby aspartate amino acid, giving rise to the positive carboxylic acid C=O absorbance at 1,764 cm^{-1}. The magnitude of the 1,764 cm^{-1} band decreases at successively more negative applied potentials, implying an electric field-induced shift in pK_a and increase in energy barrier for protonation of the aspartate at lower transmembrane potentials.

Time-resolved transmission IR approaches to study light-triggered proton reduction by nickel-iron hydrogenases have been developed by Dyer and co-workers, using either direct or mediated electron transfer [53]. Direct electron transfer was achieved by decorating the surface of the hydrogenase with either CdTe quantum dots or a $Ru(bpy)_3^{2+}$ photosensitizer (Fig. 14a). Illumination resulted in reductive activation of the active site (Fig. 14b) and concomitant proton reduction, quantified by gas chromatography (Fig. 14c). Although this is not an electrochemical measurement in the strictest sense, because there is no direct measure of the photocatalytic current, the use of gas chromatography to quantify proton reduction activity provides similar information and so this should be considered as a companion approach to the biophotoelectrochemical study of fast light-triggered electrochemical reductions.

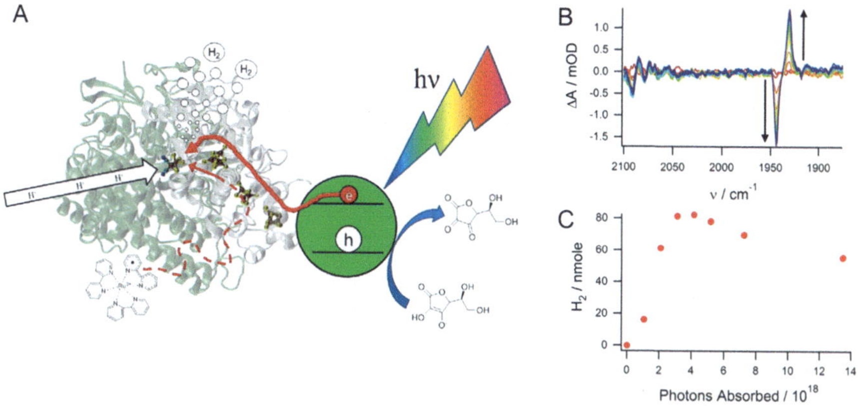

Fig. 14 Biophotoelectrochemical proton reduction by nickel-iron hydrogenases. (**a**) Schematic representation of direct electron transfer from CdTe quantum dots (*solid red arrow*) or Ru photosensitizer (*dashed red arrow*) to a nickel-iron hydrogenase active site upon illumination. (**b**) Reductive activation of the hydrogenase active site upon illumination, monitored by infrared spectroscopy. (**c**) Quantification of proton reduction using gas chromatography. Adapted with permission from Greene et al. [52]. Copyright (2012) American Chemical Society

4.5 *Spectroelectrochemical Studies of Enzymes Under Electrocatalytic Turnover*

The mechanism of H_2 oxidation by *E. coli* nickel-iron hydrogenase I (Hyd-1) has been studied by Hidalgo et al. under steady-state conditions using an ATR-IR spectroscopic method, PFIRE, at a high surface area carbon electrode [33]. As shown in Fig. 15a, the active site coordination environment of the iron in nickel-iron hydrogenases includes a carbon monoxide ligand and two cyanide ligands. The spectra shown include only the spectral region associated with the CO stretching band. Although the Fe is known to remain in the formal oxidation state of Fe^{II} throughout the catalytic cycle and oxidative inactivation side reactions of nickel-iron hydrogenases, the wavenumber position of the CO vibrational band shifts significantly in response to changes in oxidation state at the Ni or changes in the other ligands coordinating at the active site. *E. coli* Hyd-1 is poor at reducing protons, so there is no catalysis observed in the absence of H_2: see Fig. 15b in which the current under an inert Ar atmosphere remains zero at all applied potentials. Spectra recorded under Ar (Fig. 15c) therefore report on how the active site responds to the electrode potential in the absence of catalysis, giving a redox titration of the active site. At the most positive potential (+0.356 V vs SHE), the enzyme resides entirely in an oxidised inactive state known as Ni-B. At more negative potentials a range of more reduced states of the active site are generated, consistent with states previously characterised for other nickel-iron hydrogenases in solution [19]. Under an H_2 atmosphere, the enzyme exhibits an electrocatalytic H_2 oxidation current at potentials more positive than about −0.3 V vs SHE (Fig. 15d). In spectra recorded during turnover under H_2 (Fig. 15e), active site states that

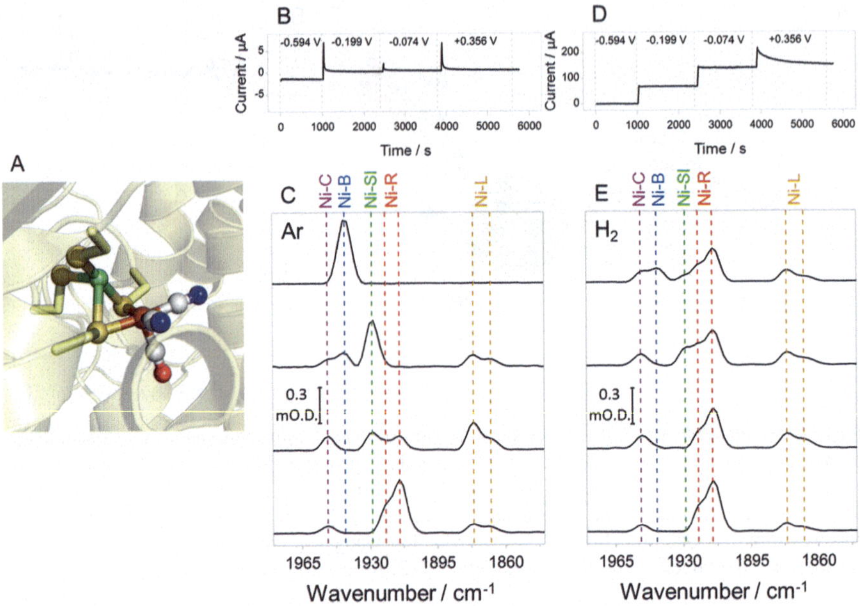

Fig. 15 Protein film IR electrochemistry (PFIRE) applied to a nickel-iron hydrogenase. (**a**) The bimetallic hydrogenase active site showing endogenous CO and CN^- ligands on iron. (**b**) *Current-time trace* showing the absence of catalytic activity under an Ar atmosphere. (**c**) Spectra in the CO stretching region, recorded in situ at each potential. (**d**) *Current-time trace* showing electrocatalytic hydrogen oxidation in the presence of H_2. (**e**) Steady-state kinetic spectra in the CO stretching region, recorded in situ. Peak assignments in (**c**) and (**e**) refer to states of the active site as discussed in the main text. Reproduced with permission from Hidalgo et al. [33]. Copyright 2015 The Authors. Published by Wiley-VCH Verlag GmbH & Co. KGaA, Weinheim

appear only under H_2, and not under Ar, should be indicative of forms of the active site involved in the catalytic cycle. The most pronounced differences are observed at the highest applied potential, +0.356 V. Of the states implicated in catalysis, Ni–C, Ni–SI and Ni–R had already been assigned as catalytic intermediates on the basis of solution studies. However, the catalytic relevance of Ni–L had been largely overlooked. The appearance of Ni–L in the PFIRE experiments under H_2 provided strong evidence for its involvement in the catalytic cycle of *E. coli* Hyd-1, and a separate study of its pH dependence by Murphy et al. assigned a pH-dependent equilibrium between Ni–C and Ni–L states (Fig. 16), providing insight into the movement of protons at the active site during catalysis [54]. Electrochemically-triggered activation of another nickel iron hydrogenase, isolated from *Desulfovibrio vulgaris* Miyazaki F, has been studied by SEIRA spectroscopy at a gold electrode under turnover (H_2) and non-turnover (Ar) conditions [55]. The potential dependence of the transition between the oxidised inactive Ni–B state and the active Ni–SI state was different under turnover and non-turnover conditions, highlighting the importance of carrying out measurements under catalytically relevant conditions.

Vibrational spectroscopy was also coupled with direct electrochemistry by Kielb et al. in a SEIRA and SERRs study of cellobiose dehydrogenase [56]. Here the

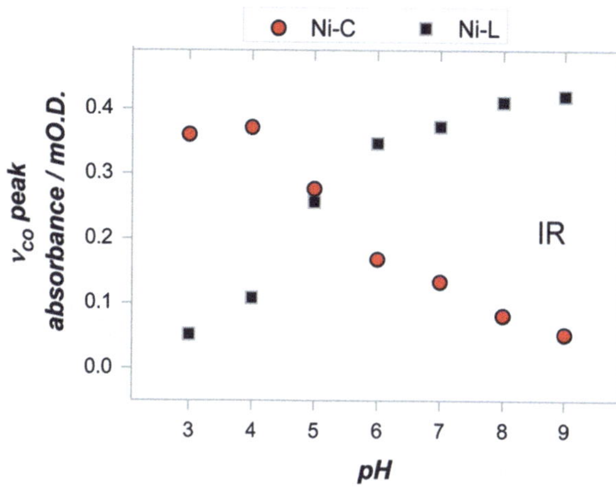

Fig. 16 Buffer exchange in the PFIRE cell, showing a pH equilibrium between the Ni–C and Ni–L states of *E. coli* Hyd-1. Reproduced with permission from Murphy et al. [54]

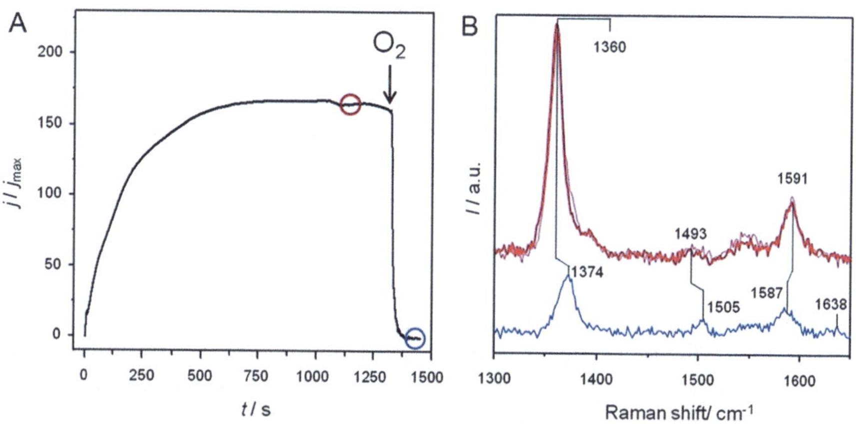

Fig. 17 The effect of O_2 on the electrocatalytic activity of an electrode modified with whole cells. (**a**) *Current-time trace* showing the suppression of electrocatalytic acetate oxidation in the presence of O_2. (**b**) In situ resonance Raman spectra recorded before (*upper trace*, *red*) and after (*lower trace*, *blue*) addition of O_2. The spectra in (**b**) were recorded at the times indicated by *circles* in (**a**). Electrocatalytic activity was recovered after removal of O_2, resulting in an almost indistinguishable Raman spectrum (**b**, *thin red line*). Reproduced from Millo et al. [57] with permission from The Royal Society of Chemistry

authors focussed on the amide region of the spectrum, and a change in signal intensity in this spectral region in response to addition of Ca^{2+} ions to solution provided information on Ca^{2+}-induced rearrangement of the haem domain in the protein. Millo et al. applied resonance Raman spectroscopy to an electrode modified with whole cells, of great relevance to microbial fuel cells, to monitor changes in the redox state of haems in electron relay multihaem proteins [57]. Figure 17a shows a current–time trace recorded during steady-state electrocatalytic oxidation of the metabolic substrate acetate. The electrocatalytic current was suppressed by addition of O_2, and resonance Raman spectra recorded in situ (Fig. 17b) revealed that this sensitivity was caused by oxidation of haem groups in the presence of O_2.

5 Related Approaches and Future Prospects

There are a number of other approaches sensitive to vibrations within biomolecules. Nuclear resonance vibrational spectroscopy (NRVS) provides information on chemistry occurring at specific, isotopically labelled, metal sites within proteins and has been used to study the reactions of NO with model iron–sulphur clusters [58] and to provide evidence for a bridging hydride in the most reduced catalytically active state, Ni–R, of nickel-iron hydrogenase active sites [59]. It is difficult to combine NRVS with direct electrochemistry, however, because of the long spectral acquisition times and the need to collect spectra at cryogenic temperatures. Studies using NRVS require relatively high protein concentrations, and are limited to metals with Mössbauer-active isotopes (including ^{57}Fe and ^{61}Ni) and so samples can be costly to produce. Sum frequency generation (SFG) spectroscopy is an intrinsically surface sensitive vibrational spectroscopic technique which has been used to study non-biological electrochemical systems [60, 61], and has also been used to study changes in protein secondary structure and orientation upon adsorption at a variety of interfaces [62–64]. It should therefore also be possible to study bioelectrodes using SFG spectroscopy. Other IR and Raman methods, such as microscopy, nano-FTIR [65] and tip-enhanced Raman spectroscopy [66, 67], could be used to bring the sensitivity of combined spectroscopic and electrochemical measurements towards the single molecule level.

The ability to gather electrochemical and spectroscopic data from the same electrode is important for the successful application of vibrational spectroscopy to bioelectrochemical systems. This allows direct correlation of the activity or rate of an electrochemical process to structural changes within a protein, providing direct spectroscopic 'snapshots' of protein function. The ability to probe coordination at metal centres, protonation changes in amino acid side chains, reaction-induced changes in organic cofactors or substrates, and subtle changes in protein secondary structure simultaneously, rapidly and at room temperature means that vibrational spectroscopic approaches are almost uniquely applicable to answering a wide range of questions in bioelectrochemistry.

Acknowledgements The authors are grateful to the European Research Council (EnergyBioCatalysis-ERC-2010-StG-258600), the Biotechnology and Biological Sciences Research Council (BB/L009722/1), and the Engineering and Physical Sciences Research Council (EP/N013514/1) for funding. We wish to thank Pathinan Paengnakorn and Charlotte McKenna for recording the IR spectrum of carboxymyoglobin in Fig. 1, Rebecca Shutt for acquiring the spectra used in Fig. 4 and Ricardo Hidalgo for experimental data collection for and preparation of Figs. 7, 15 and 16.

References

1. Siebert F, Hildebrandt P (2008) Vibrational spectroscopy in life science. Wiley-VCH, Weinheim, Chichester
2. Ash PA, Vincent KA (2012) Spectroscopic analysis of immobilised redox enzymes under direct electrochemical control. Chem Commun 48:1400–1409. doi:10.1039/C1CC15871F
3. Sezer M, Millo D, Weidinger IM et al (2012) Analyzing the catalytic processes of immobilized redox enzymes by vibrational spectroscopies. IUBMB Life 64:455–464. doi:10.1002/iub.1020
4. Atkins PW, Friedman RS (2010) Molecular quantum mechanics, 5th edn. Oxford University Press, Oxford
5. Hollas M (2004) Modern spectroscopy, 4th edn. Wiley, Chichester
6. Hecht E (2016) Optics, 5th edn. Addison-Wesley, Boston
7. Caughey WS, Shimada H, Choc MG, Tucker MP (1981) Dynamic protein structures: infrared evidence for four discrete rapidly interconverting conformers at the carbon monoxide binding site of bovine heart myoglobin. Proc Natl Acad Sci U S A 78:2903–2907
8. Peterson ES, Friedman JM, Chien EYT, Sligar SG (1998) Functional implications of the proximal hydrogen-bonding network in myoglobin: a resonance Raman and kinetic study of Leu89, Ser92, His97, and F-helix swap mutants. Biochemistry (Mosc) 37:12301–12319. doi:10.1021/bi980752u
9. Baker MJ, Trevisan J, Bassan P et al (2014) Using Fourier transform IR spectroscopy to analyze biological materials. Nat Protoc 9:1771–1791. doi:10.1038/nprot.2014.110
10. Barth A, Zscherp C (2002) What vibrations tell about proteins. Q Rev Biophys 35:369–430. doi:10.1017/S0033583502003815
11. Barth A (2007) Infrared spectroscopy of proteins. Biochim Biophys Acta 1767:1073–1101. doi:10.1016/j.bbabio.2007.06.004
12. Barth A (2000) The infrared absorption of amino acid side chains. Prog Biophys Mol Biol 74:141–173. doi:10.1016/S0079-6107(00)00021-3
13. Venyaminov SY, Prendergast FG (1997) Water (H2O and D2O) molar absorptivity in the 1,000–4,000 cm − 1range and quantitative infrared spectroscopy of aqueous solutions. Anal Biochem 248:234–245. doi:10.1006/abio.1997.2136
14. George J, Thomas J (1999) Raman spectroscopy of protein and nucleic acid assemblies. Annu Rev Biophys Biomol Struct 28:1–27. doi:10.1146/annurev.biophys.28.1.1
15. Shreve AP, Cherepy NJ, Mathies RA (1992) Effective rejection of fluorescence interference in raman spectroscopy using a shifted excitation difference technique. Appl Spectrosc 46:707–711
16. Matousek P, Towrie M, Ma C et al (2001) Fluorescence suppression in resonance Raman spectroscopy using a high-performance picosecond Kerr gate. J Raman Spectrosc 32:983–988. doi:10.1002/jrs.784
17. Hildebrandt P, Stockburger M (1984) Surface-enhanced resonance Raman spectroscopy of Rhodamine 6G adsorbed on colloidal silver. J Phys Chem 88:5935–5944. doi:10.1021/j150668a038
18. Bard AJ, Faulkner LR (2001) Electrochemical methods: fundamentals and applications, 2nd edn. Wiley, New York
19. Lubitz W, Ogata H, Rüdiger O, Reijerse E (2014) Hydrogenases. Chem Rev. doi:10.1021/cr4005814
20. De Lacey AL, Fernández VM, Rousset M, Cammack R (2007) Activation and inactivation of hydrogenase function and the catalytic cycle: spectroelectrochemical studies. Chem Rev 107:4304–4330. doi:10.1021/cr0501947
21. Moss D, Nabedryk E, Breton J, Mäntele W (1990) Redox-linked conformational changes in proteins detected by a combination of infrared spectroscopy and protein electrochemistry. Eur J Biochem 187:565–572. doi:10.1111/j.1432-1033.1990.tb15338.x
22. McEvoy JP, Foord JS (2005) Direct electrochemistry of blue copper proteins at boron-doped diamond electrodes. Electrochim Acta 50:2933–2941. doi:10.1016/j.electacta.2004.11.043

23. Renault C, Harris KD, Brett MJ et al (2011) Time-resolved UV-visible spectroelectrochemistry using transparent 3D-mesoporous nanocrystalline ITO electrodes. Chem Commun 47:1863–1865. doi:10.1039/C0CC04154H
24. Renault C, Nicole L, Sanchez C et al (2015) Unraveling the charge transfer/electron transport in mesoporous semiconductive TiO2 films by voltabsorptometry. Phys Chem Chem Phys 17:10592–10607. doi:10.1039/C5CP00023H
25. Marritt SJ, Kemp GL, Xiaoe L et al (2008) Spectroelectrochemical characterization of a pentaheme cytochrome in solution and as electrocatalytically active films on nanocrystalline metal-oxide electrodes. J Am Chem Soc 130:8588–8589. doi:10.1021/ja802641a
26. Kemp GL, Marritt SJ, Xiaoe L et al (2009) Opportunities for mesoporous nanocrystalline SnO2 electrodes in kinetic and catalytic analyses of redox proteins. Biochem Soc Trans 37:368–372. doi:10.1042/BST0370368
27. Matousek P, Parker AW (2006) Bulk Raman analysis of pharmaceutical tablets. Appl Spectrosc 60:1353–1357. doi:10.1366/000370206779321463
28. Matousek P, Stone N (2007) Prospects for the diagnosis of breast cancer by noninvasive probing of calcifications using transmission Raman spectroscopy. J Biomed Opt 12:024008–024008–8. doi: 10.1117/1.2718934
29. Leitch JJ, Brosseau CL, Roscoe SG et al (2013) Electrochemical and PM-IRRAS characterization of cholera toxin binding at a model biological membrane. Langmuir 29:965–976. doi:10.1021/la304939k
30. Olejnik P, Palys B, Kowalczyk A, Nowicka AM (2012) Orientation of laccase on charged surfaces. Mediatorless oxygen reduction on amino- and carboxyl-ended ethylphenyl groups. J Phys Chem C 116:25911–25918. doi:10.1021/jp3098654
31. Woods DA, Bain CD (2014) Total internal reflection spectroscopy for studying. Soft Matter 10:1071–1096. doi:10.1039/C3SM52817K
32. Harrick NJ (1967) Internal reflection spectroscopy. Wiley, New York
33. Hidalgo R, Ash PA, Healy AJ, Vincent KA (2015) Infrared spectroscopy during electrocatalytic turnover reveals the Ni–L active site state during H2 oxidation by a NiFe hydrogenase. Angew Chem Int Ed 54:7110–7113. doi:10.1002/anie.201502338
34. Ataka K, Stripp ST, Heberle J (2013) Surface-enhanced infrared absorption spectroscopy (SEIRAS) to probe monolayers of membrane proteins. Biochim Biophys Acta Biomembr 1828:2283–2293. doi:10.1016/j.bbamem.2013.04.026
35. Aroca R (2006) Surface-enhanced vibrational spectroscopy. Wiley, Chichester
36. Schlücker S (2014) Surface-enhanced Raman spectroscopy: concepts and chemical applications. Angew Chem Int Ed 53:4756–4795. doi:10.1002/anie.201205748
37. Gutiérrez-Sanz O, Rüdiger O, De Lacey A (2014) FTIR spectroscopy of metalloproteins. In: Nicolet Y, Fontecilla-Camps JC (eds) Metalloproteins. Humana, New York, pp 95–106
38. Królikowska A (2013) Surface-enhanced resonance Raman scattering (SERRS) as a tool for the studies of electron transfer proteins attached to biomimetic surfaces: case of cytochrome *c*. Electrochim Acta 111:952–995. doi:10.1016/j.electacta.2013.08.140
39. Ataka K, Heberle J (2007) Biochemical applications of surface-enhanced infrared absorption spectroscopy. Anal Bioanal Chem 388:47–54. doi:10.1007/s00216-006-1071-4
40. Jeuken LJC (2016) Biophotoelectrochemistry: from bioelectrochemistry to photosynthesis. Adv Biochem Eng Biotechnol. Springer, Chapter 2
41. Kriegel S, Uchida T, Osawa M et al (2014) Biomimetic environment to study *E. coli* complex I through surface-enhanced IR absorption spectroscopy. Biochemistry (Mosc) 53:6340–6347. doi:10.1021/bi500955a
42. Kranich A, Ly HK, Hildebrandt P, Murgida DH (2008) Direct observation of the gating step in protein electron transfer: electric-field-controlled protein dynamics. J Am Chem Soc 130:9844–9848. doi:10.1021/ja8016895
43. Block H, Maertens B, Spriestersbach A, Brinker N, Kubicek J, Fabis R, Labahn J, Schäfer F (2009) Immobilized-metal affinity chromatography (IMAC): a review. Method Enzymol 463:439–473. doi:10.1016/S0076-6879(09)63027-5

44. Ataka K, Giess F, Knoll W et al (2004) Oriented attachment and membrane reconstitution of His-tagged Cytochrome *c* oxidase to a gold electrode: in situ monitoring by surface-enhanced infrared absorption spectroscopy. J Am Chem Soc 126:16199–16206. doi:10.1021/ja045951h
45. Guo H, Kimura T, Furutani Y (2013) Distortion of the amide-I and -II bands of an α-helical membrane protein, pharaonis halorhodopsin, depends on thickness of gold films utilized for surface-enhanced infrared absorption spectroscopy. Chem Phys 419:8–16. doi:10.1016/j.chemphys.2012.11.011
46. Ataka K, Richter B, Heberle J (2006) Orientational control of the physiological reaction of cytochrome *c* oxidase tethered to a gold electrode. J Phys Chem B 110:9339–9347. doi:10.1021/jp0534131
47. Sezer M, Frielingsdorf S, Millo D et al (2011) Role of the HoxZ subunit in the electron transfer pathway of the membrane-bound [NiFe]-hydrogenase from Ralstonia eutropha immobilized on electrodes. J Phys Chem B 115:10368–10374. doi:10.1021/jp204665r
48. Kozuch J, Weichbrodt C, Millo D et al (2014) Voltage-dependent structural changes of the membrane-bound anion channel hVDAC1 probed by SEIRA and electrochemical impedance spectroscopy. Phys Chem Chem Phys 16:9546–9555. doi:10.1039/C4CP00167B
49. Yamakata A, Shimizu H, Oiki S (2015) Surface-enhanced IR absorption spectroscopy of the KcsA potassium channel upon application of an electric field. Phys Chem Chem Phys 17:21104–21111. doi:10.1039/C5CP02681D
50. Alves A, Ly HK, Hildebrandt P et al (2015) Nature of the surface-exposed cytochrome–electrode interactions in electroactive biofilms of desulfuromonas acetoxidans. J Phys Chem B 119:7968–7974. doi:10.1021/acs.jpcb.5b03419
51. Moe E, Sezer M, Hildebrandt P, Todorovic S (2015) Surface enhanced vibrational spectroscopic evidence for an alternative DNA-independent redox activation of endonuclease III. Chem Commun 51:3255–3257. doi:10.1039/C4CC09498K
52. Jiang X, Zaitseva E, Schmidt M et al (2008) Resolving voltage-dependent structural changes of a membrane photoreceptor by surface-enhanced IR difference spectroscopy. Proc Natl Acad Sci U S A 105:12113–12117. doi:10.1073/pnas.0802289105
53. Greene BL, Joseph CA, Maroney MJ, Dyer RB (2012) Direct evidence of active-site reduction and photodriven catalysis in sensitized hydrogenase assemblies. J Am Chem Soc 134:11108–11111. doi:10.1021/ja3042367
54. Murphy BJ, Hidalgo R, Roessler MM et al (2015) Discovery of dark pH-dependent H+ migration in a [NiFe]-hydrogenase and its mechanistic relevance: mobilizing the hydrido ligand of the Ni–C intermediate. J Am Chem Soc 137:8484–8489. doi:10.1021/jacs.5b03182
55. Millo D, Hildebrandt P, Pandelia ME et al (2011) SEIRA spectroscopy of the electrochemical activation of an immobilized [NiFe] hydrogenase under turnover and non-turnover conditions. Angew Chem Int Ed 50:2632–2634. doi:10.1002/anie.201006646
56. Kielb P, Sezer M, Katz S et al (2015) Spectroscopic observation of calcium-induced reorientation of cellobiose dehydrogenase immobilized on electrodes and its effect on electrocatalytic activity. ChemPhysChem 16:1960–1968. doi:10.1002/cphc.201500112
57. Millo D, Ly HK (2015) Towards the understanding of the effect of oxygen on the electrocatalytic activity of microbial biofilms: a spectroelectrochemical study. RSC Adv 5:92042–92044. doi:10.1039/C5RA17429E
58. Tonzetich ZJ, Wang H, Mitra D et al (2010) Identification of protein-bound dinitrosyl iron complexes by nuclear resonance vibrational spectroscopy. J Am Chem Soc 132:6914–6916. doi:10.1021/ja101002f
59. Ogata H, Krämer T, Wang H et al (2015) Hydride bridge in [NiFe]-hydrogenase observed by nuclear resonance vibrational spectroscopy. Nat Commun 6:7890. doi:10.1038/ncomms8890
60. Bozzini B, De Gaudenzi GP, Busson B et al (2010) In situ spectroelectrochemical measurements during the electro-oxidation of ethanol on WC-supported Pt-black, based on sum-frequency generation spectroscopy. J Power Sources 195:4119–4123. doi:10.1016/j.jpowsour.2010.01.017

61. Guyot-Sionnest P (2005) The mature years of sum-frequency generation are ahead. Surf Sci 585:1–2. doi:10.1016/j.susc.2005.04.021
62. Liu Y, Jasensky J, Chen Z (2012) Molecular interactions of proteins and peptides at interfaces studied by sum frequency generation vibrational spectroscopy. Langmuir 28:2113–2121. doi:10.1021/la203823t
63. Cheng H, Nikolic-Hughes I, Wang JH et al (2002) Environmental effects on phosphoryl group bonding probed by vibrational spectroscopy: implications for understanding phosphoryl transfer and enzymatic catalysis. J Am Chem Soc 124:11295–11306. doi:10.1021/ja026481z
64. Yan ECY, Wang Z, Fu L (2015) Proteins at interfaces probed by chiral vibrational sum frequency generation spectroscopy. J Phys Chem B 119:2769–2785. doi:10.1021/jp508926e
65. Amenabar I, Poly S, Nuansing W et al (2013) Structural analysis and mapping of individual protein complexes by infrared nanospectroscopy. Nat Commun 4:2890. doi:10.1038/ncomms3890
66. Zeng ZC, Huang SC, Wu DY et al (2015) Electrochemical tip-enhanced Raman spectroscopy. J Am Chem Soc 137:11928–11931. doi:10.1021/jacs.5b08143
67. Kurouski D, Mattei M, Van Duyne RP (2015) Probing redox reactions at the nanoscale with electrochemical tip-enhanced Raman spectroscopy. Nano Lett 15:7956–7962. doi:10.1021/acs.nanolett.5b04177

Adv Biochem Eng Biotechnol (2017) 158: 111–136
DOI: 10.1007/10_2016_7

Published online: 31 July 2016

Biophotoelectrochemistry of Photosynthetic Proteins

Nicolas Plumeré and Marc M. Nowaczyk

Abstract This chapter presents biophotoelectrochemical systems where one of nature's photosynthetic proteins, such as photosystem 1 (PS1), photosystem 2 (PS2), or bacterial reaction centers, are employed to create devices for technological applications. We use recent advances in biophotoelectrodes for energy conversion and sensing to illustrate the fundamental approaches in half-cell design and characterization. The aim is to guide electrochemists and photosynthetic researchers in the development of hybrid systems interfacing photosynthetic proteins with electrodes ranging from biosensors to biophotovoltaic cells. The first part gives an overview of the photosynthetic electron transfer chain with details on photosynthetic proteins and on the properties relevant for technological applications. The second part describes and critically discusses the main applications of biophotoelectrochemical cells based on photosynthetic proteins and exposes the respective requirement in electrode design. The following and final parts present the standard methodologies for the characterization of the biophotoelectrochemical half-cells with the main objectives of enhancing our mechanistic understanding of electron transfer, charge recombination, overpotential in photocurrent generation and protein degradation processes in devices, and thus open the perspectives for novel biophotoelectrochemical concepts and their rational optimization toward practical efficiencies.

Keywords Bacterial reaction centers, Biophotoelectrochemical cells, Biophotovoltaics, Biosensors, Charge carrier, Charge recombination, Overpotential, Photocurrent, Photosystem 1, Photosystem 2, Water splitting

---

N. Plumeré (✉) and M.M. Nowaczyk
Ruhr-University Bochum, Bochum, Germany
e-mail: nicolas.plumere@rub.de; http://www.ruhr-uni-bochum.de/ces/Nicolas_eng.html; http://homepage.rub.de/marc.m.nowaczyk/

Contents

1 The Photosynthetic Electron Transfer Chain and the Photosynthetic Proteins

1.1 Photosynthetic Z-Scheme

Photosynthesis, the conversion of solar to chemical energy, is one of the most fundamental processes, as photosynthetic organisms generate the energy gradient necessary to maintain and proliferate life on Earth. In particular, the invention of oxygenic photosynthesis approx. 2.5 billion years ago by photosynthetic microorganisms changed life dramatically [1]. They exploited water as an unlimited source of electrons and thereby initiated the water–oxygen cycle. Protons, electrons, and oxygen are released by light-driven oxidation of water, where electrons are temporarily stored in NADPH and subsequently used for the reduction of carbon dioxide. The potential gap between water and NADPH is bridged by the photosynthetic electron transport chain (PET) which is powered by two light-induced charge separations at two large membrane protein complexes, PS1 and PS2 (Fig. 1a), which are embedded in the thylakoid membrane. PS2 is the only enzyme that catalyzes the light-driven oxidation of water, whereby protons are released to the thylakoid lumen. The electrons are transferred to plastoquinone, a diffusible membrane-bound mediator that donates the electrons to the cytochrome b_6f complex. Simultaneously, protons are transported from the cytoplasm to the thylakoid lumen by protonation and deprotonation of plastoquinone or plastohydroquinone, respectively. Plastocyanin or cyt c_6, soluble proteins located in the thylakoid lumen,

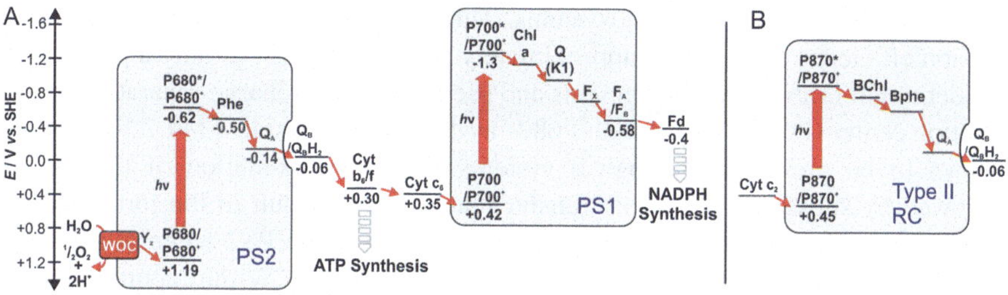

Fig. 1 The photosynthetic proteins. (**a**) The photosynthetic Z-scheme coupling the two light-induced charge separation steps at photosystem 2 and photosystem 1. (**b**) The electron transfer chain in type II bacterial reaction centers

mediate electron transfer between cytochrome b_6f complex and PS1. Finally, after the second light-induced charge separation at PS1, the electrons are transferred via ferredoxin and ferredoxin-$NADP^+$ reductase (FNR) to $NADP^+$ which is reduced to NADPH. The formation of ATP from ADP and phosphate is catalyzed by ATP-synthase which is driven by the proton gradient between the cytoplasm and the thylakoid lumen.

1.2 Photosynthetic Proteins

The separation of the two processes (1) light harvesting and (2) charge separation is a general design principle of natural photosynthesis. Light energy is harvested by antennae proteins (LHCs, phycobilisomes, intrinsic antennae) and transferred via Förster resonance energy transfer to the reaction centers (RCs), where primary charge separation is induced. This open structure enables the flexible adjustment of the light harvesting apparatus depending on the light conditions and, as a subject of the pigment composition of the antennae proteins, also light harvesting with an increased spectral cross section. In particular, the packing of pigments is critical for the efficiency, as chlorophyll–chlorophyll distances less than 10 Å promote self-quenching and therefore dissipation of excitation energy, although chlorosomes, the disordered antennae complexes of green-sulfur bacteria, follow a different design principle which is not fully understood so far [2]. Nevertheless, separation of light harvesting and charge separation requires a well-defined coupling between the two functional elements. Whereas excitation transfer between the light harvesting pigments is extremely fast (~femtoseconds), the final transfer to the charge separating pigments is slower (~picoseconds) because of a relatively longer distance. This configuration ensures efficient 'trapping' of light energy by the reaction centers, as the charge separation process is faster than the unwanted back transfer of excitation energy.

Optimization of these primary processes during evolution generated a light-harvesting-reaction-center unit with a remarkable quantum yield of nearly 100%

(photon-to-charge) for the photosystems. This corresponds to a solar energy conversion efficiency (η_{SOLAR}) of approximately 34% for PS2 [3] as only a part of the incident light is absorbed by pigments and used to generate charge separation at the reaction center ($P680^{+}P680^{*}$), with $P680^{+}$ as the strongest oxidant (+1.2 V vs SHE) known to be present in biological systems (Fig. 1a). Excitation of the RC is followed by subsequent electron transfer reactions that result in the formation of specific radical pairs by reduction of redox cofactors at the PS2 acceptor side and concurrent oxidation of redox cofactors at the donor side. Within approximately 3 ps the electron is transferred to the primary acceptor pheophytin ($P680^{+}Phe^{-}$). This is followed by electron transfer within 300 ps to plastoquinone A ($P680^{+}Q_A^{-}$). Oxidation of a nearby tyrosine residue (TyrZ) leads to reduction of $P680^{+}$ in about 100 ns ($TyrZ^{+}Q_A^{-}$). This radical pair is stable for milliseconds, which fits with the time scale of water oxidation (~2 ms) and electron transfer to plastoquinone B (~0.2 ms for oxidized Q_B, ~0.6 ms for Q_B^{-} and ~2 ms for an empty Q_B binding side [4]). $TyrZ^{+}$ abstracts electrons from the CaO_5Mn_4 cluster that stores four electrons, derived from water oxidation. Finally, Q_B is reduced and after receiving a second electron the process becomes quasi-irreversible by protonation and the formation of QH_2. The long distance between the charges (~5 nm) and the energy barrier that is raised with each step (loss of energy: ~50%) stabilizes charge recombination sufficiently to create a time window for the subsequent reactions. Taking into account that release of oxygen within 2 ms is the rate-limiting step in PS2 catalysis, a theoretical turnover frequency (TOF) of 500 e^{-} s^{-1} can be calculated. Practically, oxygen evolving activities of up to 5,000–6,000 μmol O_2 μg Chl^{-1} h^{-1} have been reported for isolated PSII complexes [5] which correspond to a TOF of 200 e^{-} s^{-1}, but the PS2 lifetime is very short (minutes) under strong illumination, which results in a relatively low turnover number (TON) of about 10^4. In contrast, the TOF in vivo is lower (85 e^{-} s^{-1}) [6], but as the lifetime is longer (~1 h) [7], a TON of about 10^5 is reached.

The short lifetime of isolated PS2 is the major challenge in any application of nature's water splitting catalyst. Nevertheless, the integration of natural photosynthetic water splitting in hybrid devices can provide insights for understanding the process itself and inspiration for alternative water oxidizing photoanodes concepts. For instance, two recent concepts for biophotovoltaic cells based on PS2 proposed the use of the H_2O/O_2 redox couple as charge carrier [8, 9] (see Sect. 2.2). Moreover, the principle of PS2-based photoanodes is opening novel sensing concepts as demonstrated for herbicide detection (see Sect. 2.1).

The quantum yield of the primary processes and also the loss of energy caused by electron transfer across the membrane and for stabilization of charge separation within PS1 are comparable to those values of PS2. Excitation of P700, the RC of PS1, leads to charge separation and formation of the first radical pair ($P700^{+}P700^{*}$), with $P700^{*}$ as the strongest reductant (−1.3 V vs SHE) known to be present in biological catalysts (Fig. 1a). Subsequently, the electron is transferred within ~1 ps to chlorophyll A_0 (Chl a) and it reaches phylloquinone A_1 (Q or vitamin K1) within ~30 ps. Electron transfer to the first (F_x) and subsequent iron sulfur clusters (F_A/F_B) is, at ~200 ns and ~500 ns, considerably slower, but the radical pair $P700^{+}F_B^{-}$ is stabilized for 60 ms to allow docking and electron transfer to the diffusible mediator ferredoxin (Fd) and re-reduction of $P700^+$ by plastocyanine or cytochrome c_6. Hence, in contrast to PS2, PS1 is not catalytically active but instead serves as an "electron pump," inducing a charge separation of 1 V corresponding to the difference in potential between P700 and F_B. In the native system, PS1 electron transfer (TOF: 47 $e^- s^{-1}$) is limited by PS2 electron donation and the PS1/PS2 ratio [6], but the higher stability of PS1 (~40 h) [7] leads to a much higher TON (~7×10^6). The rate-limiting step ($P700^+F_B^-$ formation within ~500 ns) allows theoretically a TOF in the range of 10^6 $e^- s^{-1}$. This extreme electron transfer rate, together with the high charge separation and strong reducing force, are the most compelling properties of PS1 toward applications. A recent example at the single protein level demonstrates that PS1 under extreme illumination can deliver a charge separation of 1 V at electron transfer rates in excess of 2 millions s^{-1} [10], which correlates with the intrinsic properties of native PS1. In principle, if this concept could be extended to a large ensemble of PS1 complexes, biophotoelectrochemical devices could outperform systems based on semiconductor materials. Moreover, PS1 delivers electrons energetic enough for proton or CO_2 reduction in water. Hence, the coupling of the F_B site in PS1 to redox catalysts can in principle be exploited for the production of solar fuels. This light to chemical or light to electrical energy conversion is the main motivation behind the development of novel biophotoelectrode concepts based on PS1. Their feasibility and potential applicability are supported by the reasonable robustness of PS1 and the possibility for its production on a large scale from algae.

The structure of type II reaction centers of anoxygenic photosynthetic bacteria (e.g., purple bacteria such as *Rhodobacter sphaeroides*) is similar to PS2, with the fundamental difference that the water-splitting unit is missing (Fig. 1b). Instead, electrons are donated to the complex via cytochrome c_2 in the native system. The resulting light-driven cyclic electron flow via the reaction center (RC), the quinone pool, the bc_1 complex, and Cyt c_2 generates a proton gradient via the membrane used for ATP formation. Light is harvested by the peripheral antenna complexes (LH2) and transferred to the core antenna (LH1) that surrounds the RC. Electrons are transferred after charge separation at a dimer of bacteriochlorophyll molecules (P870) via bacteriochlorophyll (B), bacteriopheophytin (H), and menaquinone A (Q_A) to ubiquinone B (Q_B), and the electron transfer times [11] are comparable to those of PSII (see above). The rate-limiting step (formation of $P^+Q_B^-$ within ~100 μs) adjusts the charge separation to the subsequent processes (docking of diffusible mediators, protonation of Q_B). The calculated TOF is about one order of magnitude higher compared to PS2 but, in principle, direct removal of electrons from Q_A would result in a one million factor increase in the theoretical TOF, as the electron transfer is not limited by water oxidation. Because the bacterial reaction centers lack catalytic activity, they primarily serve as electron pumps in analogy to PS1, and hence open up similar perspectives. Although the light-induced charge separation of the native bacterial reaction centers is lower than for PS1, the genetic engineering is straightforward and hence represents a promising building block both for sensing and energy conversion applications.

2 Applications

Photosynthetic proteins are envisioned to play a role in a variety of technological devices such as biosensors, photoelectrochemical cells for solar fuel generation, or biophotovoltaic cells for electricity production. Regardless of the application, electrodes interfaced with photosynthetic proteins can be built as a photoanode or as a photocathode. Numerous electronic contacting (direct or mediated electron transfer) and immobilization strategies have been proposed and reviewed elsewhere [12, 13]. In the following sections we discuss the main requirements for the design of the devices for sensing and energy conversion applications.

2.1 Biosensors

In biosensors based on photosynthetic proteins, the photocurrent correlates with the concentration of an analyte. The classical examples are photoanodes based on PS2 or photocathodes based on bacterial reaction centers (Fig. 2) used as herbicide biosensors. Molecules such as terbutryn or atrazine bind in the Q_B pocket of PS2 and block the electron transfer pathway associated with the light-induced water oxidation reaction [15]. Thus, the inhibition of PS2 in the photoanode decreases the photocurrent and its calibration vs the concentration of the herbicide opens up the possibility for their quantification. Because of the low photostability of PS2, research efforts have focused on the use of whole cells instead of isolated photosystems [16]. An alternative approach to circumvent the fragility of PS2 consists in using the more robust type II bacterial reaction centers because their Q_B binding sites are similar to the one in PS2. The selectivity toward a specific herbicide can be optimized via genetic engineering of the Q_B binding site both in PS2 [17] and in bacterial reaction centers [14].

In contrast to classical amperometric biosensors built on redox enzymes, the sensing concept based on photosynthetic proteins is less prone to electrochemical interferences because redox active species present in the sample matrix generate a dark current that serves as a background for the analyte dependent photocurrent. Beyond the inhibition-based devices, photoelectrochemical cells may also find applications for direct analyte sensing by coupling a reductase as a recognition element to the reducing end of a photosynthetic protein such as PS1. In this case, the analyte would serve as the final electron acceptor and the photocurrent intensity would correlate with the analyte concentration according to Michaelis–Menten behavior defined by the reductase. Again, comparison between dark and light currents enables the subtraction of Faradaic contributions produced by redox-active interferences which react at the underlying electrode. Such a concept is yet to be demonstrated whereby the main hurdle resides in the difficulty of directly coupling a protein complex such as PS1 and a reductase via their respective redox sites.

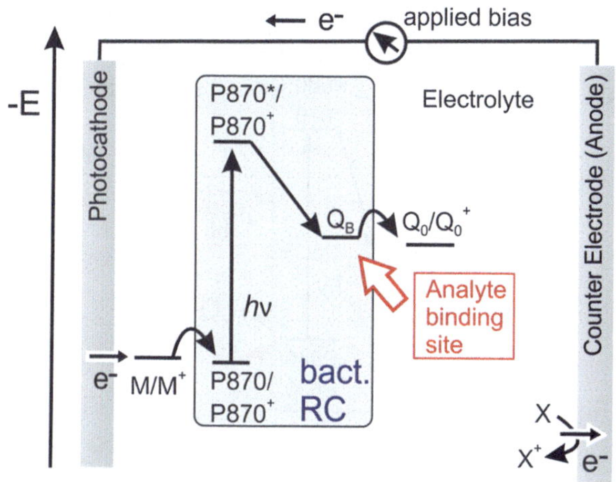

Fig. 2 Biosensor based on bacterial reaction centers (RC) in a photocathode configuration. The binding of the analyte blocks the electron transfer between ubiquinone (Q_0/Q_0^+) which serves as electron acceptor and thus decreases the photocurrent. The sensing performances are defined by the photoelectrode half-cell alone. The potentiostat controls the counter electrode potential and the resulting reaction ($X \rightarrow X^+ + e$) may involve reoxidation of Q_0 or H_2O oxidation. In this particular example, cytochrome C serves as electron mediator (M/M^+) in the photocathode half-cell [14]

The advantages of photosynthetic-based biosensors with respect to redox-active interferences open up the possibility for their technological application in point-of-care and field sensing. To achieve this goal, the main objectives in the sensor design are (1) to engineer the proteins toward the desired selectivity for a specific analyte, (2) to achieve efficient electronic contact between the proteins and the electrode to generate high photocurrent for high sensitivity in analyte detection, and (3) to design an immobilization strategy to stabilize the protein for both storage and operational stability.

2.2 *Biophotoelectrochemical Water Splitting*

Utilization of the light-induced water splitting reaction is envisioned as one of the main strategies for sustainable and clean energy production. In principle, the implementation of photosynthetic proteins and of H_2-evolving catalysts in a photoelectrochemical cell could enable this light to chemical energy conversion [12]: PS2 catalyzes the splitting of water into O_2, protons, and electrons, and the subsequent light-induced charge separation at PS1 provides enough energy to the electron for the generation of NADPH (see Sect. 1.1). This electron is also energetic enough for the evolution of hydrogen provided that a suitable catalyst is coupled to PS1.

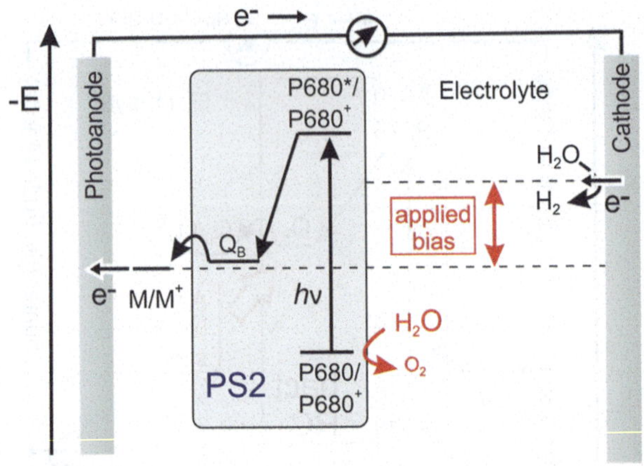

Fig. 3 Biophotoelectrochemical cell based on a photosystem 2 photoanode associated with a hydrogen-evolving electrode for light-induced water splitting under an applied bias. DCBQ can be used as electron mediator (M/M^+) between the photoanode and PS2 [19]

Until now, full biophotoelectrochemical water splitting relying exclusively on light as the source of energy and on photosynthetic proteins as the photoactive materials has not been realized. Nevertheless, the first steps toward such semi-artificial photosynthetic Z-scheme (Fig. 1a) were proposed recently. In a first report, the association of a PS2 modified photoanode with a PS1 modified photocathode demonstrated that, under light illumination, electrons can be shuttled between the two electrodes without any applied bias nor sacrificial electron donor or acceptor [9, 18]. It was even possible to regain energy from light in the form of electricity (see Sect. 2.3). However, because PS1 was not associated with a hydrogen-evolving catalyst, full water splitting could not be demonstrated. In a subsequent and complementary report, a PS2-modified photoanode for light-induced O_2 evolution was coupled to a hydrogenase-modified cathode for H_2 evolution with a light-to-hydrogen conversion efficiency of 5.4% [19]. In this case, the light-driven H_2 evolution required an additional energy input in the form of an external bias to compensate for the insufficient reducing force of the electron delivered by PS2 (Fig. 3). The integration of a second light-induced charge separation step in analogy to the natural photosynthetic Z-scheme (Fig. 1a) is expected to open up the possibility for full biological water splitting relying on light energy only. Such a biophotocathode could be constructed from PS1 associated with an H_2-evolving catalyst such as a hydrogenase [20] or Pt nanoparticles [21]. Optimization of the biophotocathodes with respect to overpotential and electron transfer rate (see Sect. 4) opens up the possibility for efficient solar fuel production. However, the intrinsic fragility of PS2 renders the future applicability of such systems uncertain.

2.3 Biophotovoltaics

Photovoltaic devices convert light to electricity by exploiting semiconductor materials or molecular photosensitizers as photoactive components. Although photovoltaic technology is already implemented on a global scale, the search for cheaper and less energy-demanding materials for device fabrication is ongoing. The photosynthetic proteins emerged as one class of alternative materials for light-induced charge separation. Their extreme quantum efficiency and large natural abundances make them ideal candidates for integration in biohybrid photovoltaic devices [22]. Although photosystems and bacterial reaction centers have evolved to limit intra-protein charge recombination (see Sect. 1.2), the major issue to consider is how to avoid charge recombination of the charge carriers when they are integrated in photovoltaic devices.

In contrast to the electron mediator (M/M^+) that transfers electrons within the half-cell, the charge carrier (Y/Y^+, see Fig. 4) is defined as the redox couple that transfers the charge between anode and cathode. For effective power generation, the charge produced at the photosynthetic protein must not be quenched at the photoelectrode it originates from, and instead should be transferred to the counter electrode to close the circuit. The difference in (over)potential for the charge transfer between the photosynthetic electrode and the (over)potential for the reaction of the charge carrier at the collector electrode defines the open circuit voltage (OCV).

Only a few complete biophotovoltaic cells based on photosynthetic proteins have been reported until now. The first generation was based on isolated bacterial reaction centers incorporated in semiconducting electron transport layers to interface the biomolecules with the electrodes [23]. However, the extreme illumination intensity (as high as 100 suns) used in this study could induce major photocurrent contributions from photocorrosion processes, and so achieving long-term stabilities under such conditions may not be feasible. A later photovoltaic cell was built from

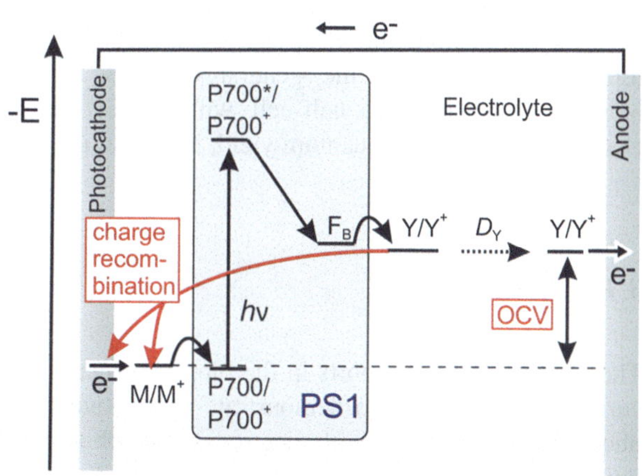

Fig. 4 Biophotovoltaic cell based on photosystem 1 in a photocathode configuration. The driving force for charge recombination (*red arrows*) increases with the difference in potential between the electron mediator (M/M^+) and the charge carrier (Y/Y^+)

PS1 interfaced with semiconductor materials developed for dye-sensitized solar cell (DSSC) technology. In DSSC, the semiconductor materials are designed to limit charge recombination processes with the charge carrier. The same principles were applied for the construction of the PS1-semiconductor hybrid photoanode [24]. A peptide blocking layer and a semiconducting electrode material were used to limit the charge recombination of an electrochemically reversible redox couple based on a Co-complex serving as charge carrier. Standard cell characterization under 1 sun illumination validated this hybrid photovoltaic cell concept with an open circuit voltage of 0.5 V and a short circuit current in excess of 300 $\mu A\ cm^{-2}$. Several other examples of semiconductor-photosynthetic protein hybrid systems [25] were reported following this initial work. However, the use of semiconductor materials that have intrinsic photoactive properties attenuates the advantage of implementing a photoactive biological component.

The first example of a semiconductor-free biophotovoltaic cell was built from a PS2 photoanode combined with an oxygen reducing cathode [8]. Light-induced charge separation at the PS2 photoanode triggers water oxidation. The produced oxygen diffuses to the cathode where it is reduced back to water via an enzyme catalyzed process. Although chemically reversible, the H_2O/O_2 redox couple that serves as the charge carrier is electrochemically irreversible, and thus impedes charge recombination (see Sect. 4.3) at the PS2 photoanode. The use of H_2O/O_2 as charge carrier was also proposed in the biophotovoltaic cell coupling a PS2 photoanode to a PS1 photocathode [9, 18]. In this approach, O_2 is again exploited as the charge carrier. The additional light-induced charge separation step at PS1 opens the possibility for coupling a catalyst for solar fuel generation [21] (see Sect. 2.2).

3 Half-Cell Characterization

Irrespective of the intended application, in-depth characterization of the respective half-cell is essential for a rational optimization of the complete device. In the following, we discuss the general methods for the characterization of the biophotoelectrochemical half-cell with respect to photocurrent, overpotential, charge recombination, quantum yield, and stability.

3.1 Photocurrent

The vast majority of reports in the biophotoelectrochemical field focus on photocurrent measurements to demonstrate conceptually the possibility of interfacing photosynthetic proteins with electrodes. The photocurrent, in analogy to the catalytic current obtained from electrodes modified with redox enzymes, is related to the activity of the photosynthetic protein and to the various electron transfer processes associated with the light-induced charge separation.

The first step in the characterization of a biophotoelectrode is to measure the photocurrent and to identify its source. Besides the charge separation at the photosynthetic protein, other processes such as (1) photocorrosion, (2) light-induced temperature changes, or (3) photoactivity of the underlying electrode material can induce background photocurrents. Validation of the role of the photosynthetic protein is often performed by investigating the effect of a specific inhibitor on the photocurrent generation. Numerous herbicides such as dinoterb (2,4-dinitro-6-*tert*-butylphenol) are able to block the natural electron transfer from the Q_B site of the D1 subunit in photosystem 2. Hence, the quenching of photocurrent generation upon addition of such inhibitors was proposed as a control experiment to demonstrate the role of the photosystem. However, several reports have shown that, upon integration in an artificial system, the natural electron transfer chain can be altered whereby the electron can exit the PS2 directly from Q_A (Fig. 1a) and hence bypasses Q_B, the inhibitor binding site. For instance, when PS2 is directly contacted to ITO electrodes [26], significant photocurrents assigned to PS2 are observed even in the presence of the inhibitor. Hence, the use of inhibitors for the native electron transfer chain as control experiments to identify the source of photocurrent may lead to misinterpretation and to the oversight of alternative and potentially valuable electron transfer pathways in hybrid systems which interface photosynthetic proteins and electrode materials.

Instead of inhibitors, a more reliable and direct approach to assign unambiguously the contributions to photocurrent generation is to record the photocurrent as a function of the wavelength of the incident light. The photocurrent action spectrum of the biophotoelectrodes should coincide with the absorption spectrum of the isolated protein. This approach was employed in recent reports on PS1 [27, 28], PS2 [29, 30], and bacterial reaction centers [31, 32] (Fig. 5). The action spectra are advantageous because they directly reveal the photocurrent contribution by each component of the biophotoelectrode. This investigation is essential when intrinsically photoactive semiconductors are used as electrode materials for the immobilization of the photosynthetic proteins.

The intensity of the photocurrent depends on the protein loading, and its determination is necessary for the subsequent electrode optimization. The presence of numerous chlorophylls in the photosystems (35 in PS2 and about 100 in the PS1 from *Thermosynechococcus elongatus*) facilitates the quantification of the electrode surface coverage by protein via absorbance measurements. In the case of transparent electrode material, the visible light spectrum of the complete electrode is recorded and the protein loading is determined from the chlorophyll absorbance. The strategy was applied for the quantification of PS1 [28, 33] and PS2 [30]. If the electrodes are non-transparent, the chlorophylls can be extracted with organic solvents such as acetone or methanol and quantified via their absorbance in solution to deliver indirectly the amount of protein initially present on the electrodes [19, 31]. In the case of monolayers on flat electrodes, the loading is too low for quantification via the chlorophylls. Instead, surface plasmon resonance [29, 34],

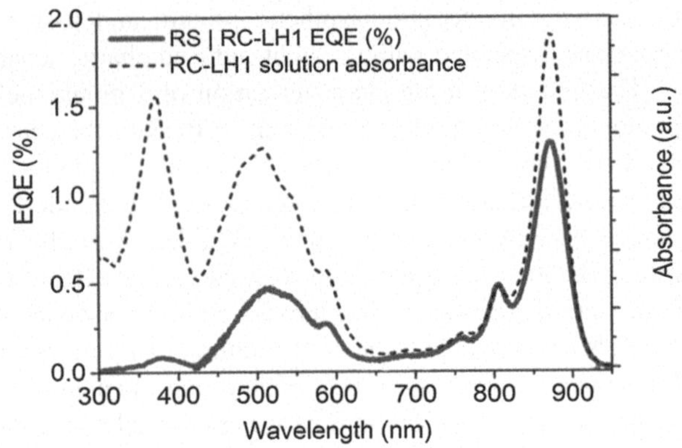

Fig. 5 Absorbance spectrum (*dotted line*) vs action spectrum (*solid line*) given as the wavelength dependent external quantum efficiency (EQE) of bacterial reaction center associated with the light harvesting complex (RC-LH1) on rough silver electrodes (RS). Adapted from [31]

atomic force microscopy [35], and, in principle, quartz crystal microbalance [36] can be applied to determine the surface coverage.

Knowledge of the protein content is essential to normalize the photocurrent as the number of electrons flowing through PS1 every second, or in other words, the turnover frequency (TOF). The TOF values facilitate the comparison of the performances and the identification of rate-limiting steps toward subsequent optimization. For instance, a recent example of PS1 photocathodes based on an Os-complex modified hydrogel [33] resulted in a TOF in excess of 300 e s^{-1}. This value is the highest reported to date for PS1 (similar values were achieved for bacterial reaction centers on Ag electrodes [31]) and is significantly larger than the ones observed in vivo. Nevertheless, it remains several orders of magnitude below the theoretical maximum based on the time scale of the electron transfer within PS1 (see Sect. 1.2). The main objective with respect to photocurrent in the design of PS1- (or bacterial reaction center-) based photoelectrodes is to achieve non-limiting electronic contact of the protein to the electrode to exploit its properties fully. Once electron transfer rates are maximized, the associated photocurrent can be scaled up by increasing the protein loading on the electrode (see Sect. 4.1). Although challenging, this strategy could open up the possibility for biophotoelectrochemical devices competing with or even outperforming systems based on semiconducting materials.

To gain further insights into the rate-limiting process for photocurrent generation, the determination of the kinetics of the individual electron transfer steps is desired. In the case of classical redox enzyme modified electrodes, the extraction of rate constants from enzymatic catalytic currents [37], obtained for example from cyclic voltammetry, is well established both for mediated [38] and direct electron transfer [39]. However, the application of this procedure to photosynthetic proteins is not straightforward because charge recombination processes (see below) compete with the electron transfer chain for photocurrent generation. Nevertheless, if specific experimental conditions can be found for which charge recombination becomes insignificant, the kinetic constants related to the reactions between natural as well as artificial electron mediators and the photosynthetic proteins can be determined. This was demonstrated for PS1 in solution with both natural and artificial redox partners [40].

In the direct electron transfer configuration of the photosystem on the electrode, the interfacial electron transfer rate can in principle be determined in the dark, and hence independently of charge recombination. If the redox centers of the photosynthetic proteins can be detected electrochemically in the dark, the determination of the heterogeneous electron transfer rate becomes accessible via the classical Laviron methodology. For instance, the immobilization of PS1 [41] as well as PS2 [42] on highly oriented pyrolytic graphic electrodes revealed redox signals that could be assigned to redox sites of the photosystems. Although the interfacial electron transfer rates could be extracted, the subsequent illumination of the electrode did not reveal any photocurrents, indicating that the immobilized proteins were not in their native states. On the other hand, PS2 immobilized in a DET configuration on meso-ITO materials displays significant photocurrent under illumination [19]. However, the large background currents induced by the porous ITO materials did not enable the detection of the Q_B or Q_A redox sites in the dark, and hence impeded the determination of the heterogeneous electron transfer rate between the native PS2 protein complexes and the electrode.

Reports of individual electron transfer steps in biophotoelectrodes are mostly absent in the literature. However, such a systematic methodology is necessary to optimize the highly complex processes involved in photocurrent generation and to use the photosynthetic proteins to their full potential.

3.2 Onset Potential/Overpotential for Photocurrent Generation

Characterization of the onset potential for photocurrent generation is essential for energy conversion applications because it defines, together with the photocurrent density, the power output and the overall energy conversion efficiency. The onset potential depends on the overpotential for electron transfer between the electrode and the photosynthetic protein (see [37]). A straightforward approach for its determination is to record a cyclic voltammogram [43] under illumination at low scan rate [29, 33]. For maximized energy conversion efficiency, the onset potential for photocurrent generation needs to approach the redox potential of the respective redox site in the photosynthetic protein.

In analogy to classical electrocatalytic processes [44], the onset potential or overpotential in photocurrent generation is quantified from the photocurrent wave response obtained under non-limiting substrate or charge carrier mass transport. The half-wave potential used for overpotential determination is accurately extracted by taking the derivative of the photocurrent response in the cyclic voltammogram. For instance, the half-wave potential for a bacterial reaction center based photocathode coincides with the redox potential of the cytochrome C used as

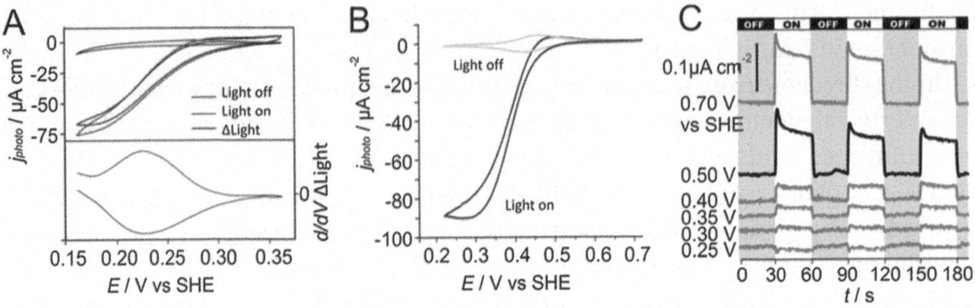

Fig. 6 Determination of the onset potential for photocurrent generation: (**a**) by means of cyclic voltammetry for a bacterial reaction center based photocathode (adapted from [31]); (**b**) by means of cyclic voltammetry for a PS1 based photocathode (adapted from [33]); (**c**) by means of potential step voltammetry for a PS2 based photoanode (adapted from [26])

electron relay and hence validates its role in interfacing the photosynthetic protein with the electrode [31] (Fig. 6a). In the case of PS1 integrated in a poly(vinyl) imidazole Os(bispyridine)$_2$Cl polymer matrix, the photocurrent displays a half-wave potential of about 380 mV vs SHE, which is only 40 mV negative to the redox potential of the P_{700} redox center in PS1 [33] (Fig. 6b), hence demonstrating photocurrent generation at low overpotential.

If the time scale of the experiment is excessively long compared to the stability of the biophotoelectrodes, potential step voltammetry [43] may be applied instead of cyclic voltammetry to reduce the overall illumination time needed for the determination of the onset potential for photocurrent generation. In the case of extreme instability, a fresh electrode can be prepared for each potential step. In a recent report on PS2 in a DET configuration on meso-ITO electrodes, the potential step method revealed photocurrents at an onset potential of about 300 mV vs SHE which is 360 mV positive to the Q_B site in PS2 [26] (Fig. 6c).

3.3 Charge Recombination

The major difference between bioelectrochemistry and biophotoelectrochemistry lies in the possible direction of the electron flow for a given electrochemical process. In classical bioelectrocatalysis, the charge transfer follows a single direction between the electrode and the substrate according to the thermodynamic driving force. In contrast, in biophotoelectrochemistry, the light-induced process injects energy into the electron transfer chain, which opens up the possibility for dual electron transfer pathways of opposite directions. The high energy electron generated at the photosynthetic protein can react with a subsequent redox partner such as a charge carrier or a final electron acceptor. However, a second and often favored process is the recombination of this electron with an oxidized charge carrier or the electrode from which it was generated (red arrows, Fig. 4). This charge recombination process cancels parts or all of the photocurrent and is often

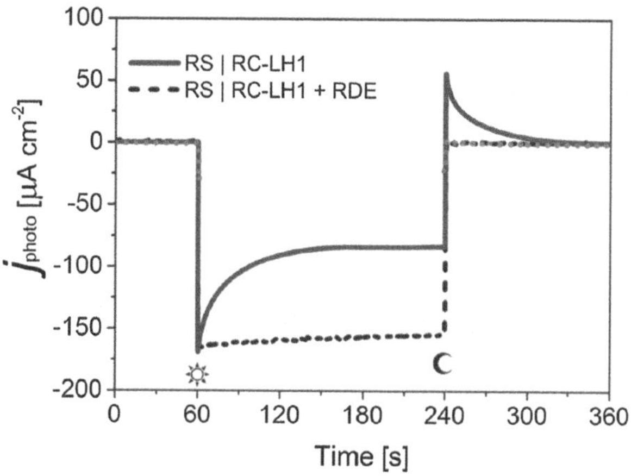

Fig. 7 Photocurrent density produced by bacterial reaction center associated with the light harvesting complex (RC-LH1) on rough silver electrodes (RS) under stationary conditions (*solid line*) and under rotation (RDE, *dashed line*). Adapted from [31]

unnoticed in photoelectrochemical systems. Diagnostic tools to identify charge recombination and strategies to bypass this process are needed for the development of efficient technological devices.

The issue related to the recombination of charge carriers was already recognized in the first biophotoelectrochemical measurements with PS1 in solution [40, 45]. Identification of the charge recombination is straightforward if its kinetics are moderate. In potential step voltammetry, the charge recombination results in a reversal of the current when switching from illumination to dark conditions. For instance, in the case of a photocathode, the reduced charge carrier produced during illumination can be reoxidized at the electrode surface and hence decreases the overall cathodic photocurrent. Upon switching the light source off, this reoxidation process persists for a time scale depending on the charge recombination kinetics as well as on the mass transport of the charge carrier and generates a net anodic current. This effect can be verified by using forced convection to transport the reduced charge carrier away from the photoelectrode surface (Fig. 7). In contrast to the stationary conditions, the photocurrent is higher and the dark current vanishes [31].

When the kinetics of charge recombination are fast, the photocurrents are completely quenched and the current response resembles the case of an inactive photoelectrode. Electrochemical methods for the quantification of this process remain to be developed. The coupling of electrochemical and laser methods may be advantageous for this purpose.

3.4 *Quantum Efficiency*

The external quantum efficiency (EQE) or incident photon-to-current conversion efficiency (IPCE) is the ratio of the flux of electrons involved in the net photocurrent to the flux of incident photons [31, 46]. The internal quantum efficiency (IQE)

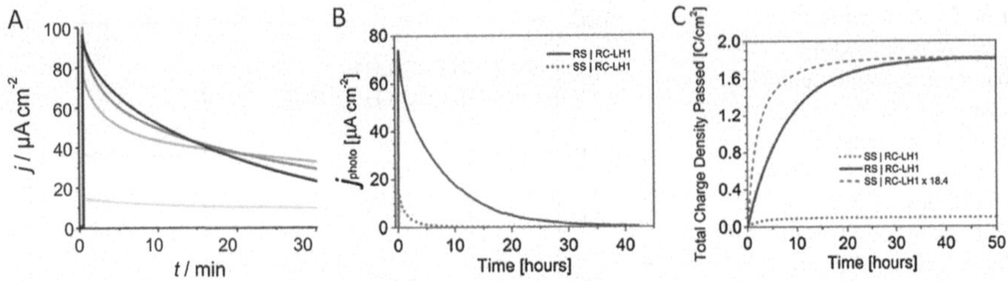

Fig. 8 Stability determination. (**a**) Photocurrent decay for a photocathode based on PS1 in an Os-complex modified hydrogel under various illumination intensities (*from black to light gray*: 40, 25, 10, and 1 mW cm^{-2} at 685 nm). Adapted from [33]. (**b**) Photocurrent from a photocathode based on a bacterial reaction center on smooth silver (SS) and rough silver (RS) electrode. (**c**) The associated TON expressed as the total charged passed (adapted from [31])

is the ratio of the number of transferred electrons to the number of photons absorbed by the photoelectrode. The IQE is calculated by taking the number of electrons per second obtained from the photocurrent and the number of absorbed photons per second obtained from the absorbance of the photoelectrode and the light illumination intensity [28].

A low IQE can be caused by one or several of the following processes: (1) extensive charge recombination, (2) poor electronic connection of the photosynthetic proteins, (3) an insufficient amount of photosynthetic proteins or (4) components of the biophotoelectrode acting as a black body converting the absorbed light into heat.

The quantum efficiency is often mistaken for energy conversion efficiency. Although both numbers are expressed as ratios, they have different meanings. The quantum efficiency does not account for the energy of the incident light or for the energy of the transferred electrons and hence does not provide any insight for the overall energy conversion efficiency. The EQE and IQE can be calculated for half cells because the energy involved in driving the reaction at the counter electrode is not relevant. On the other hand, the energy conversion efficiency can only be determined for the complete cell because the energy involved in every step including the counter electrode reaction must be accounted for (see Sect. 4).

3.5 Stability Determination

The stability of biophotoelectrodes is a key parameter in evaluating their applicability. The loss of activity of a photosynthetic protein is often expressed as its half-life for given intensities and wavelengths of the incident light. A recent PS1 photocathode [33] displayed half-lives for the photocurrent at a constant applied potential ranging from below 30 min for the strongest irradiation (40 mW cm^{-2} at 685 nm) up to 15 h at 1 mW cm^{-2} (Fig. 8a). Because light illumination not only affects the half-lives but also the absolute value of the generated photocurrent, a more insightful approach to characterize the photoelectrode stability is to report the turn over number (TON) for a given set of illumination conditions. The TON of a photoelectrode is the total charge obtained by integrating the photocurrent over time until complete loss of activity [31] (Fig. 8b, c).

4 Cell Optimization

The performance of biophotoelectrochemical cells still require significant improvements before technological applications can be considered. For instance, in biophotovoltaic devices the overall efficiency (η) in light to electricity conversion is calculated from the integral photocurrent density (i_{photo}), the open-circuit photovoltage (V_{oc}), the fill factor of the cell (*ff*), and the intensity of the incident light (I_s):

$$\eta = i_{\text{photo}} \times V_{\text{oc}} \times ff / I_s$$

To date, the energy conversion efficiencies of devices that implement biological components for light-induced charge separation have not exceeded 0.1%, which is two orders of magnitude below that of the standard semiconductor photovoltaics or of the dye-sensitized solar cell analogues [47, 48]. Finding solutions to achieve practical efficiency and stability with biophotoelectrochemical devices requires a fundamental reconsideration of the present bio-photoelectrode concepts and the rational optimization of their individual components. It is essential that all targeted properties of the bio-photoelectrodes are considered with respect to energy conversion efficiency and that the individual components are engineered accordingly. In the following, we present the main strategies (1) to increase photocurrent density, (2) to circumvent charge recombination, (3) to maximize the open circuit voltage, and (4) to improve the stability of the devices.

4.1 Photocurrent, Protein Loading, and Electron Transfer

Photocurrents are defined by (1) the activity of the photosynthetic proteins, (2) their loading, and (3) the various electron transfer steps between electrodes, electron mediators, charge carriers and proteins (see Sect. 3.1). Increasing the loading in photosynthetic proteins is the most straightforward approach to improve the photocurrent densities significantly. To this end, high-surface-area electrodes such as porous plasmonic silver [31] or inverse opals made of meso-ITO [19] were used to set benchmarks for photocurrent densities with bacterial reaction centers and PS2, respectively. However, the use of a porous electrode implies the need for transparent or plasmonic materials that might be excessively costly for large scale applications. The use of polymeric matrices for the immobilization of large amounts of photosynthetic proteins on flat electrodes could serve as an advantageous alternative because, in this case, transparency of the underlying electrode is not necessary and therefore low cost material becomes suitable. For instance, PS1 complexes were immobilized and electronically contacted via redox hydrogels on planar carbon based electrodes [33] achieving photocurrents competing with the above-mentioned porous electrodes systems. Although transparency of such polymeric films could be achieved by proper selection of its building blocks (polymer backbone and redox mediators), the electron transfer properties within the matrix can benefit from further development aiming at non-limiting electronic contacting of the photosynthetic protein to the electrode.

To maximize photocurrents it is also necessary to optimize the kinetics of the electron transfer between the photosynthetic protein and the electrode. In the DET configuration, the orientation of the protein with its respective redox sites close to the electrode surface is essential. The charge of the electrode surface can be tuned to achieve a preferential and desired orientation of the protein. For instance, a negatively charged electrode surface results in an electrostatic immobilization and orientation of PS2 with its Q_B site facing the electrode. Alternatively, site-directed mutagenesis is a versatile tool to introduce or replace specific amino acid residues in a protein to favor an isotropic orientation. Cysteine residues can be used to orient the protein via disulfide or self-assembly on Au, and polyhistidine tags (6–10 histidine residues) interact specifically with nitrilotriacetic acid (NTA)-coated surfaces via immobilized metal ions (e.g., nickel). To date, the use of these amino acid residues for controlled orientation only had limited success in achieving fast direct electron transfer under electrochemical conditions.

When electron mediators are employed, selection of the proper redox moiety for fast electron transfer with the redox protein is critical. In the case where the electron mediators are immobilized within a polymeric matrix together with the protein, it is important that the electron transfer between these electron relays is fast as well. This is typically achieved with small distances between the tethered redox mediators (i.e., high concentration) and a high mobility/solvation of the matrix [33].

4.2 Overpotential and Open Circuit Voltage

High energy conversion efficiency requires electron transfer at low overpotential to minimize energy losses. For instance, in a biophotovoltaic cell the onset potential for photocurrent generation of the photoelectrode half-cell and the overpotential for the electron transfer between charge carrier and the collector electrode define the open circuit voltage (Fig. 4). In the mediated electron transfer configuration, the redox potential of the electron mediator needs to be adjusted to that of the redox center in the protein. For instance, in a photocathode based on PS1 integrated with a redox hydrogel, an Os complex serves as electron donor for the P_{700} center in PS1. The potential of the Os complex was tuned to be only 25 mV negative to P_{700} to induce a slight driving force for fast electron transfer at minimized overpotential [33]. A similar strategy was applied to contact PS2 to a photoanode via toluidine blue as electron mediator [18]. In the DET case, overpotential is associated with the kinetics of the electron transfer with the electrode (see [37]). Hence, proper orientation of the protein on the electrode to minimize the electron transfer distance with the respective redox site is beneficial. Nevertheless, efficient DET with photosynthetic protein at low overpotential remains to be demonstrated (see Sects. 3.1 and 3.2).

A second strategy to decrease the onset potential for photocurrent generation is to recover the electron earlier within the intraprotein electron transfer chain (Fig. 1a). Photosynthetic reaction centers contain optimized electron tunneling pathways to cross the membrane barrier of 35 Å and to stabilize charge separation in the millisecond time scale for electron transfer to a mobile carrier (see Sect. 1). The application of isolated proteins in an artificial environment offers the opportunity to re-engineer internal electron transfer pathways and to create novel electron outlets specifically adjusted to the artificial redox environment. In a pioneering study, PS2 was modified by a single amino acid exchange at the cytoplasmic surface near Q_A that allows docking of cytochrome c [49]. After blocking of the natural electron outlet via Q_B by the inhibitor 3-(3,4-dichlorophenyl)-1,1-dimethylurea (DCMU), the electrons are transferred via the novel artificial outlet Q_A to the soluble mediator cytochrome c. The novel electron transfer pathway is supported by the large potential difference between Q_A and cytochrome c (~500 mV) and the long stability (milliseconds) of the $TyrZ^+Q_A^-$ radical pair. This study shows the capability for PS2 engineering but the gain in potential at the level of Q_A (−140 mV vs SHE) compared to Q_B (−60 mV vs SHE) is rather low. Harnessing the electron at the more negative potential of pheophytine (−500 mV vs SHE) in PS2 would (1) allow novel applications which are usually dependent on PS1 and (2) open the possibility for novel strategies to increase the efficiency of the biophotovoltaic device, but the implementation is much more challenging.

From the biologist's perspective, any intervention in the natural function of the photosystem needs a backup strategy that ensures (1) survival of the source organism, (2) accurate assembly of multisubunit-multicofactor-protein complex, and (3) sufficient stability in the natural environment. Isolation of photosystems is a challenging task as these large membrane protein complexes – in particular PS2 – are easily damaged during solubilization and purification. Most applications rely on stable photosystems that were isolated from thermophilic cyanobacteria such as *Thermosynechococcus elongatus*. Genetic engineering is established for this organism although it is more challenging compared to other cyanobacteria. A major drawback of this protein source is its intrinsic photoautotrophy, which means that it requires functional photosystems to survive. Other cyanobacteria such as *Synechocystis* sp. PCC 6803 are easier to manipulate and can be grown in the presence of a reduced carbon source (e.g., glucose) without functional photosystems, but the protein complexes are more fragile in general compared to those of *T. elongatus*. Several groups are currently trying to engineer *T. elongatus* strains able to grow heterotrophically. However, if such a strain was available, its practical applicability for the isolation of modified PS2 complexes would need to be investigated in further studies. An alternative strategy could be selective tagging and the specific isolation of mutated photosystems. Modification of a redox center that leads to a malfunctioning or non-functioning photosystem can be coupled in principle with an affinity marker (e.g., His-tag) that can be used for the selective isolation of the engineered photosystem. However, in both approaches the modification needs to be at least compatible with the intricate assembly process and should not affect the stability of the complex. For instance, cyanobacterial PS2 is composed of 17 membrane intrinsic and 3 extrinsic proteins, as well as more than 80 cofactors per monomer that have to be assembled in a concerted process, which is assisted by numerous additional proteins [50]. Moreover, any modification that effects the forward electron transfer to Q_B may also lead to increased light-induced damage of PS2 [51], which may result in rapid degradation of the modified complex within the living organism. Although it is a challenging task, the interplay between engineered photosystems and tailor-made redox environments bears a great potential for optimization of biophotoelectrochemical devices.

4.3 Bypassing Charge Recombination

Charge recombination remains one of the major challenges to be overcome to achieve practical efficiencies with biophotoelectrochemical cells. Upon light-induced charge separation, the photosynthetic protein converts a low energy electron to a high energy electron, which is recovered by a charge carrier. Charge recombination within the photosynthetic protein is circumvented by a fast intraprotein electron transfer, which induces a spatial separation between the hole and the high energy electron (see Sect. 1). However, once the electron is transferred from the protein to the charge carrier, because of the large thermodynamic driving force, the latter preferentially recombines with the photoelectrode. In dye-sensitized solar cells, semiconductor materials are used to impede charge recombination and this strategy inspired the development of hybrid semiconductor-biophotovoltaic devices (see Sect. 2.3). In the following we discuss strategies for bypassing charge recombination in biophotoelectrodes that do not require semiconductor materials, and thus demonstrate that devices based on photosynthetic proteins as the only photoactive component can be applied for light energy conversion.

Kinetically Protected Charge Carriers

In natural photosynthesis, charge carriers such as the NADPH/$NADP^+$ redox couple kinetically protect the high energy electron from recombination: The electrons can only be exchanged with a redox partner in the presence of a catalyst (FNR, see

Sect. 1.1) and hence the undesired charge recombination, for instance with oxygen, is avoided. Nature's approach relying on a catalyzed interconversion between the reduced and oxidized form of the charge carrier may be valuable to prevent redox side-reactions in biophotoelectrochemical devices as well. By exploiting chemically reversible but electrochemically irreversible redox couples as charge carriers in combination with a reversible catalyst, the location of the electron transfer can be defined.

This concept was demonstrated in recent reports on biophotovoltaics (see Sect. 2.3) based on PS2 with H_2O/O_2 as an electrochemically irreversible charge carrier. The light-induced oxidation of water by PS2 generates O_2 and a high energy electron (about −60 mV vs SHE) which upon transfer to the electrode generates an anodic photocurrent. O_2 can in principle recombine with this electron, however, by proper selection of the electron mediator (phenothiazine dyes) in MET [18] or of the electrode material (ITO) in DET [19], although this back reaction at the photoanode can be impeded. Conversely, to enable the conversion of the charge carrier at the counter/collector electrode, in this case the cathode, the latter is modified with a catalyst for O_2 reduction such as bilirubin oxidase [8]. Hence, the need for a catalyzed charge carrier conversion ensures that the latter is reduced exclusively at the counter electrode, thus closing the electrical circuit.

Spatially Separated Electron Transfer Pathways

Natural photosynthesis, besides using the kinetically protected $NADPH/NADP^+$ redox couple as charge carrier, also benefits from the spatial separation of redox mediators via the thylakoid membrane. A similar strategy may also be applied for photoelectrode design. The directed orientation of a dense monolayer of PS1 would allow for electronic contact to a generator and to a collector electrode in a photovoltaic configuration. Here, the main strategy relies on contacting both the electron-accepting and the electron-donating sites of PS1 with surface-confined electron relays or via direct electron transfer between two electrodes. The spatial separation of the electron transfer pathway before and after light-induced charge separation could efficiently block charge recombination. The challenge resides in the cell fabrication because PS1 needs to be inserted in a nanogap between the two electrodes. The concept was shown at the single molecule level in a double junction involving direct electron transfer [10]. The current density and open circuit voltage of this bionanodevice opens the prospect for highly efficient biophotovoltaics. However, upscaling this concept to a large surface area may be difficult because defects in the nanogap would result in short-circuits between the two electrodes [52].

Spatially separated electron transfer pathways for biophotoelectrochemistry can also be built from redox proteins that are not photoactive. In a recent report, a decaheme cytochrome was used to shuttle electrons between a photoactive semiconductor particle and the electrode [53]. The electron transfer pathway is shielded by the protein matrix and hence charge recombination is circumvented. The strategy could in principle be applied in the same manner to contact a photosynthetic protein to an electrode.

Lowering the Driving Force for Charge Recombination

The kinetics of the charge recombination between charge carriers depends on the driving force imposed by their difference in potential (Fig. 4). Decreasing the driving force to limit the charge recombination is of limited interest for energy conversion because it implies a lower open circuit voltage. Nevertheless, the strategy may be of interest for biophotoelectronic applications where photocurrent generation is the main requirement, such as in biosensing (see Sect. 2.1). This possibility to limit the charge recombination is illustrated by a photoelectrochemical cell built from bacterial reaction centers with its electron-accepting side (P_{870}) interfaced to a silver electrode via a bridging cytochrome C. A benzoquinone derivative served as final electron acceptor for the photoexcited electron exiting the reaction center. Although the electrochemical reoxidation of the hydroquinone is quasireversible and should in principle result in charge recombination, the latter process is limited by the small difference in potential between the quinone and the cytochrome C (or the applied potential at the Ag electrode). As a result, very high photocurrents (>400 μA cm^{-2}) are obtained [31].

4.4 *Improving Stability*

In the development of high performance biophotoelectrodes, the stability under illumination must be considered. Therefore, concepts based on the more robust PS1 and bacterial reaction centers, compared to the fragile PS2 biomolecule, may become more widespread.

The main deactivation pathways of PS1 are related to O_2. PS1 photoinhibition is induced by light-triggered singlet-oxygen generation [54] and by the partially reduced O_2 species formed by the strongly reducing potential of the electron-exiting PS1. In artificial systems, when O_2 is used as the final electron acceptor, the partially reduced oxygen intermediates, including superoxide and H_2O_2, are suspected to be responsible for the fast deactivation of PS1 with a half-life as low as 30 min under strong illumination [33]. Therefore, a photoelectrochemical cell intended for applications should be designed to operate exclusively in an anaerobic state. Under these conditions, the stability of PS1 in vivo (about 40 h, see Sect. 1.2) is anticipated to be surpassed. The elimination of trace O_2, can be achieved with sacrificial oxygen scavenger systems [55] whereby systems already demonstrated for bioelectrochemical applications [56] are anticipated to be transposable to biophotoelectrochemical applications.

Type II reaction centers from anaerobic bacteria, in contrast, are relatively immune to O_2. In comparison to PS1, the lower energy of the electron leaving the reaction center typically does not result in significant reaction between the electron acceptor and oxygen [31]. Hence the deactivation pathways involving partially reduced oxygen species are avoided.

References

1. Hohmann-Marriott MF, Blankenship RE (2011) Evolution of photosynthesis. Annu Rev Plant Biol 62(1):515–548. doi:10.1146/annurev-arplant-042110-103811
2. Noy D, Moser CC, Dutton PL (2006) Design and engineering of photosynthetic light-harvesting and electron transfer using length, time, and energy scales. Biochim Biophys Acta 1757(2):90–105. doi:10.1016/j.bbabio.2005.11.010
3. Dau H, Zaharieva I (2009) Principles, efficiency, and blueprint character of solar-energy conversion in photosynthetic water oxidation. Acc Chem Res 42(12):1861–1870. doi:10.1021/ar900225y
4. de Wijn R, van Gorkom HJ (2002) The rate of charge recombination in photosystem II. Biochim Biophys Acta 1553(3):302–308. doi:10.1016/S0005-2728(02)00183-4
5. Kuhl H, Kruip J, Seidler A, Krieger-Liszkay A, Bunker M, Bald D, Scheidig AJ, Rögner M (2000) Towards structural determination of the water-splitting enzyme: purification, crystallization, and preliminary crystallographic studies of photosystem II from a thermophilic cyanobacterium. J Biol Chem 275(27):20652–20659. doi:10.1074/jbc.M001321200
6. Lubner CE, Applegate AM, Knörzer P, Ganago A, Bryant DA, Happe T, Golbeck JH (2011) Solar hydrogen-producing bionanodevice outperforms natural photosynthesis. Proc Natl Acad Sci U S A 108(52):20988–20991. doi:10.1073/pnas.1114660108
7. Yao DC, Brune DC, Vermaas WF (2012) Lifetimes of photosystem I and II proteins in the cyanobacterium *Synechocystis* sp. PCC 6803. FEBS Lett 586(2):169–173. doi:10.1016/j.febslet.2011.12.010
8. Yehezkeli O, Tel-Vered R, Wasserman J, Trifonov A, Michaeli D, Nechushtai R, Willner I (2012) Integrated photosystem II-based photo-bioelectrochemical cells. Nat Commun 3:742. doi:10.1038/ncomms1741
9. Kothe T, Plumeré N, Badura A, Nowaczyk MM, Guschin DA, Rögner M, Schuhmann W (2013) Combination of a photosystem 1-based photocathode and a photosystem 2-based photoanode to a Z-scheme mimic for biophotovoltaic applications. Angew Chem Int Ed 52 (52):14233–14236. doi:10.1002/anie.201303671
10. Gerster D, Reichert J, Bi H, Barth JV, Kaniber SM, Holleitner AW, Visoly-Fisher I, Sergani S, Carmeli I (2012) Photocurrent of a single photosynthetic protein. Nat Nanotechnol 7 (10):673–676. doi:10.1038/nnano.2012.165
11. Moore GF, Brudvig GW (2011) Energy conversion in photosynthesis: a paradigm for solar fuel production. Annu Rev Condens Matter Phys 2(1):303–327. doi:10.1146/annurev-conmatphys-062910-140503
12. Kothe T, Schuhmann W, Rögner M, Plumeré N (2015) 9 Semi-artificial photosynthetic Z-scheme for hydrogen production from water. In: Rögner M (ed) Biohydrogen. De Gruyter, Berlin, München, Boston
13. Tel-Vered R, Willner I (2014) Photo-bioelectrochemical cells for energy conversion, sensing, and optoelectronic applications. ChemElectroChem 1(11):1778–1797. doi:10.1002/celc.201402133
14. Swainsbury DJK, Friebe VM, Frese RN, Jones MR (2014) Evaluation of a biohybrid photoelectrochemical cell employing the purple bacterial reaction centre as a biosensor for herbicides. Biosens Bioelectron 58:172–178. doi:10.1016/j.bios.2014.02.050
15. Giardi MT, Pace E (2005) Photosynthetic proteins for technological applications. Trends Biotechnol 23(5):257–263. doi:10.1016/j.tibtech.2005.03.003

16. Lambreva MD, Giardi MT, Rambaldi I, Antonacci A, Pastorelli S, Bertalan I, Husu I, Johanningmeier U, Rea G (2013) A powerful molecular engineering tool provided efficient Chlamydomonas mutants as bio-sensing elements for herbicides detection. PLoS One 8(4), e61851. doi:10.1371/journal.pone.0061851
17. Rea G, Polticelli F, Antonacci A, Scognamiglio V, Katiyar P, Kulkarni SA, Johanningmeier U, Giardi MT (2009) Structure-based design of novel *Chlamydomonas reinhardtii* D1-D2 photosynthetic proteins for herbicide monitoring. Protein Sci 18(10):2139–2151. doi:10.1002/pro.228
18. Hartmann V, Kothe T, Pöller S, El-Mohsnawy E, Nowaczyk MM, Plumeré N, Schuhmann W, Rögner M (2014) Redox hydrogels with adjusted redox potential for improved efficiency in Z-scheme inspired biophotovoltaic cells. Phys Chem Chem Phys 16(24):11936. doi:10.1039/c4cp00380b
19. Mersch D, Lee CY, Zhang JZ, Brinkert K, Fontecilla-Camps JC, Rutherford AW, Reisner E (2015) Wiring of photosystem II to hydrogenase for photoelectrochemical water splitting. J Am Chem Soc 137(26):8541–8549. doi:10.1021/jacs.5b03737
20. Krassen H, Schwarze A, Friedrich B, Ataka K, Lenz O, Heberle J (2009) Photosynthetic hydrogen production by a hybrid complex of photosystem I and [NiFe]-hydrogenase. ACS Nano 3(12):4055–4061. doi:10.1021/nn900748j
21. Zhao F, Conzuelo F, Hartmann V, Li H, Nowaczyk MM, Plumeré N, Rögner M, Schuhmann W (2015) Light induced H2 evolution from a biophotocathode based on photosystem 1 - Pt nanoparticles complexes integrated in solvated redox polymers films. J Phys Chem B 119 (43):13726–13731. doi:10.1021/acs.jpcb.5b03511
22. Nguyen K, Bruce BD (2014) Growing green electricity: progress and strategies for use of photosystem I for sustainable photovoltaic energy conversion. Biochim Biophys Acta 1837 (9):1553–1566. doi:10.1016/j.bbabio.2013.12.013
23. Das R, Kiley PJ, Segal M, Norville J, Yu AA, Wang L, Trammell SA, Reddick LE, Kumar R, Stellacci F, Lebedev N, Schnur J, Bruce BD, Zhang S, Baldo M (2004) Integration of photosynthetic protein molecular complexes in solid-state electronic devices. Nano Lett 4 (6):1079–1083. doi:10.1021/nl049579f
24. Mershin A, Matsumoto K, Kaiser L, Yu D, Vaughn M, Nazeeruddin MK, Bruce BD, Graetzel M, Zhang S (2012) Self-assembled photosystem-I biophotovoltaics on nanostructured TiO_2 and ZnO. Sci Rep 2:234. doi:10.1038/srep00234
25. Ocakoglu K, Krupnik T, van den Bosch B, Harputlu E, Gullo MP, Olmos JDJ, Yildirimcan S, Gupta RK, Yakuphanoglu F, Barbieri A, Reek JNH, Kargul J (2014) Photosystem I-based biophotovoltaics on nanostructured hematite. Adv Funct Mater 24(47):7467–7477. doi:10.1002/adfm.201401399
26. Kato M, Cardona T, Rutherford AW, Reisner E (2013) Covalent immobilization of oriented photosystem II on a nanostructured electrode for solar water oxidation. J Am Chem Soc 135 (29):10610–10613. doi:10.1021/ja404699h
27. Yehezkeli O, Wilner OI, Tel-Vered R, Roizman-Sade D, Nechushtai R, Willner I (2010) Generation of photocurrents by bis-aniline-cross-linked Pt nanoparticle/photosystem I composites on electrodes. J Phys Chem B 114(45):14383–14388. doi:10.1021/jp100454u
28. Ciesielski PN, Faulkner CJ, Irwin MT, Gregory JM, Tolk NH, Cliffel DE, Jennings GK (2010) Enhanced photocurrent production by photosystem I multilayer assemblies. Adv Funct Mater 20(23):4048–4054. doi:10.1002/adfm.201001193
29. Badura A, Esper B, Ataka K, Grunwald C, Wöll C, Kuhlmann J, Heberle J, Rögner M (2006) Light-driven water splitting for (bio-)hydrogen production: photosystem 2 as the central part of a bioelectrochemical device. Photochem Photobiol 82(5):1385. doi:10.1562/2006-07-14-RC-969
30. Efrati A, Tel-Vered R, Michaeli D, Nechushtai R, Willner I (2013) Cytochrome *c*-coupled photosystem I and photosystem II (PSI/PSII) photo-bioelectrochemical cells. Energy Environ Sci 6(10):2950. doi:10.1039/c3ee41568f

31. Friebe VM, Delgado JD, Swainsbury DJK, Gruber JM, Chanaewa A, van Grondelle R, von Hauff E, Millo D, Jones MR, Frese RN (2016) Plasmon-enhanced photocurrent of photosynthetic pigment proteins on nanoporous silver. Adv Funct Mater 26:285–292. doi:10.1002/adfm.201504020
32. Tan SC, Crouch LI, Jones MR, Welland M (2012) Generation of alternating current in response to discontinuous illumination by photoelectrochemical cells based on photosynthetic proteins. Angew Chem Int Ed 51(27):6667–6671. doi:10.1002/anie.201200466
33. Kothe T, Pöller S, Zhao F, Fortgang P, Rögner M, Schuhmann W, Plumeré N (2014) Engineered electron-transfer chain in photosystem 1 based photocathodes outperforms electron-transfer rates in natural photosynthesis. Chem Eur J 20(35):11029–11034. doi:10.1002/chem.201402585
34. Stieger KR, Feifel SC, Lokstein H, Lisdat F (2014) Advanced unidirectional photocurrent generation via cytochrome *c* as reaction partner for directed assembly of photosystem I. Phys Chem Chem Phys 16(29):15667–15674. doi:10.1039/c4cp00935e
35. Manocchi AK, Baker DR, Pendley SS, Nguyen K, Hurley MM, Bruce BD, Sumner JJ, Lundgren CA (2013) Photocurrent generation from surface assembled photosystem I on alkanethiol modified electrodes. Langmuir 29(7):2412–2419. doi:10.1021/la304477u
36. Feifel SC, Stieger KR, Lokstein H, Lux H, Lisdat F (2015) High photocurrent generation by photosystem I on artificial interfaces composed of π-system-modified graphene. J Mater Chem A 3(23):12188–12196. doi:10.1039/C5TA00656B
37. Börner RA (2016) Isolation and cultivation of anaerobes. Adv Biochem Eng Biotechnol. doi:10.1007/10_2016_1
38. Limoges B, Marchal D, Mavré F, Savéant JM (2006) Electrochemistry of immobilized redox enzymes: kinetic characteristics of NADH oxidation catalysis at diaphorase monolayers affinity immobilized on electrodes. J Am Chem Soc 128(6):2084–2092. doi:10.1021/ja0569196
39. Heering HA, Hirst J, Armstrong FA (1998) Interpreting the catalytic voltammetry of electroactive enzymes adsorbed on electrodes. J Phys Chem B 102(35):6889–6902. doi:10.1021/jp981023r
40. Proux-Delrouyre V, Demaille C, Leibl W, Sétif P, Bottin H, Bourdillon C (2003) Electrocatalytic investigation of light-induced electron transfer between cytochrome *c*6 and photosystem I. J Am Chem Soc 125(45):13686–13692. doi:10.1021/ja0363819
41. Munge B, Das SK, Ilagan R, Pendon Z, Yang J, Frank HA, Rusling JF (2003) Electron transfer reactions of redox cofactors in spinach photosystem I reaction center protein in lipid films on electrodes. J Am Chem Soc 125(41):12457–12463. doi:10.1021/ja036671p
42. Alcantara K, Munge B, Pendon Z, Frank HA, Rusling JF (2006) Thin film voltammetry of spinach photosystem II. Proton-gated electron transfer involving the Mn 4 cluster. J Am Chem Soc 128(46):14930–14937. doi:10.1021/ja0645537
43. Faulkner R, Bard AJ (2001) Electrochemical methods: fundamentals and applications, 2nd edn. Wiley, New York
44. Artero V, Saveant JM (2014) Toward the rational benchmarking of homogeneous H_2-evolving catalysts. Energy Environ Sci 7(11):3808–3814. doi:10.1039/C4EE01709A
45. Allen H, Hill O, Walton NJ, Whitford D (1985) The coupling of heterogeneous electron transfer to photosystem 1. J Electroanal Chem Interfacial Electrochem 187(1):109–119. doi:10.1016/0368-1874(85)85579-9
46. Kato M, Zhang JZ, Paul N, Reisner E (2014) Protein film photoelectrochemistry of the water oxidation enzyme photosystem II. Chem Soc Rev 43(18):6485–6497. doi:10.1039/c4cs00031e
47. Boschloo G, Hagfeldt A (2009) Characteristics of the iodide/triiodide redox mediator in dye-sensitized solar cells. Acc Chem Res 42(11):1819–1826. doi:10.1021/ar900138m
48. Gratzel M (2001) Molecular photovoltaics that mimic photosynthesis. Pure Appl Chem 73 (3):459–467. doi:10.1351/pac200173030459

49. Larom S, Salama F, Schuster G, Adir N (2010) Engineering of an alternative electron transfer path in photosystem II. Proc Natl Acad Sci U S A 107(21):9650–9655. doi:10.1073/pnas.1000187107
50. Heinz S, Liauw P, Nickelsen J, Nowaczyk M (2016) Analysis of photosystem II biogenesis in cyanobacteria. Biochim Biophys Acta 1857(3):274–287. doi:10.1016/j.bbabio.2015.11.007
51. Vass I (2012) Molecular mechanisms of photodamage in the Photosystem II complex. Biochim Biophys Acta 1817(1):209–217. doi:10.1016/j.bbabio.2011.04.014
52. Plumeré N (2012) Single molecules: a protein in the spotlight. Nat Nanotechnol 7 (10):616–617. doi:10.1038/nnano.2012.175
53. Hwang ET, Sheikh K, Orchard KL, Hojo D, Radu V, Lee CY, Ainsworth E, Lockwood C, Gross MA, Adschiri T, Reisner E, Butt JN, Jeuken LJC (2015) A decaheme cytochrome as a molecular electron conduit in dye-sensitized photoanodes. Adv Funct Mater 25 (15):2308–2315. doi:10.1002/adfm.201404541
54. Sonoike K (1996) Photoinhibition of photosystem I: its physiological significance in the chilling sensitivity of plants. Plant Cell Physiol 37(3):239–247
55. Plumeré N (2013) Interferences from oxygen reduction reactions in bioelectroanalytical measurements: the case study of nitrate and nitrite biosensors. Anal Bioanal Chem 405 (11):3731–3738. doi:10.1007/s00216-013-6827-z
56. Plumeré N, Henig J, Campbell WH (2012) Enzyme-catalyzed O_2 removal system for electrochemical analysis under ambient air: application in an amperometric nitrate biosensor. Anal Chem 84(5):2141–2146. doi:10.1021/ac2020883

Adv Biochem Eng Biotechnol (2017) 158: 137–158
DOI: 10.1007/10_2015_5010

Published online: 18 March 2016

Artificial Photosynthesis: Hybrid Systems

Yan Ni and Frank Hollmann

Abstract Oxidoreductases are promising catalysts for organic synthesis. To sustain their catalytic cycles they require efficient supply with redox equivalents. Today classical biomimetic approaches utilizing natural electron supply chains prevail but artificial regeneration approaches bear the promise of simpler and more robust reaction schemes. Utilizing visible light can accelerate such artificial electron transport chains and even enable thermodynamically unfeasible reactions such as the use of water as reductant.

This contribution critically summarizes the current state of the art in photoredox-biocatalysis (i.e. light-driven biocatalytic oxidation and reduction reactions).

Keywords Biocatalysis, Oxidation reactions, Photocatalysis, Reduction reactions

Contents

Y. Ni and F. Hollmann (✉)
Delft University of Technology, Delft, The Netherlands
e-mail: f.hollmann@tudelft.nl

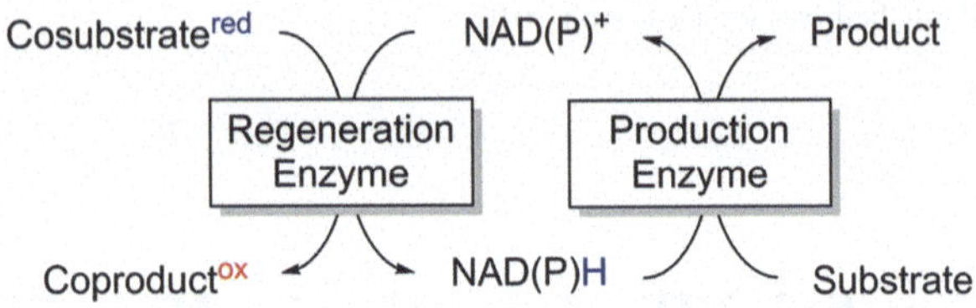

Scheme 1 General representation of established enzymatic regeneration systems for NAD(P)H-dependent production enzymes

1 Introduction

Oxidoreductases are very useful catalysts for organic synthesis as they often enable specific redox transformations at selectivities not accessible with established chemical catalysts [1, 2]. To make full use of nature's arsenal of oxidoreductases, efficient methods that sustain the oxidoreductases' catalytic cycles are vital. In other words, electrons need to be delivered to or taken away from the enzymes' active sites. Within their natural environment (generally whole living cells) this task is accomplished by cofactors that "wire" the oxidoreductase to cellular metabolism. The latter, however, has been optimized to sustain the cell's survival and function and not to provide an efficient network for preparative application of the cell. From a chemist's perspective it is generally most desirable if the oxidoreductase of interest operates at full speed which obviously is not always in line with the cell's requirements. Therefore, it is not surprising that, from an early stage on, one branch of biocatalysis research has focussed on efficient regeneration systems [3–7]. Today, a broad range of different regeneration approaches are established, some of them on an industrial scale. The majority of these approaches are biomimetic as they rely on a second enzymatic process to provide the redox equivalents needed for the oxidoreductase of interest. The redox mediator used in these systems is typically NAD(P)H (Scheme 1).

Most oxidoreductases known today rely on redox-active metals or organic molecules bound to the oxidoreductases' active sites to perform the actual reaction. Scheme 2 summarizes the most important cofactors and prosthetic groups discussed in this contribution.

In addition to these established methodologies, electrochemical and photochemical in situ regeneration approaches are also under investigation [8–11]. Today, an increased interest in photochemical methods has occurred for various reasons. Photochemical approaches in principle enable simplified reaction schemes, avoiding additional (enzymatic) regeneration catalysts. Photochemical processes frequently utilize homogeneously dissolved catalysts and thereby in principle overcome diffusion limitations often encountered with electrochemical approaches. Finally, light energy can serve as a "catalyst" to accelerate chemical reactions but also as source of energy to make thermodynamically unfavourable transformations feasible.

The aim of this chapter is to provide a tutorial overview of photochemical regeneration of cofactors and oxidoreductases and to provide a critical review of the current trends in photobiocatalysis with a focus on processes utilizing isolated oxidoreductases.

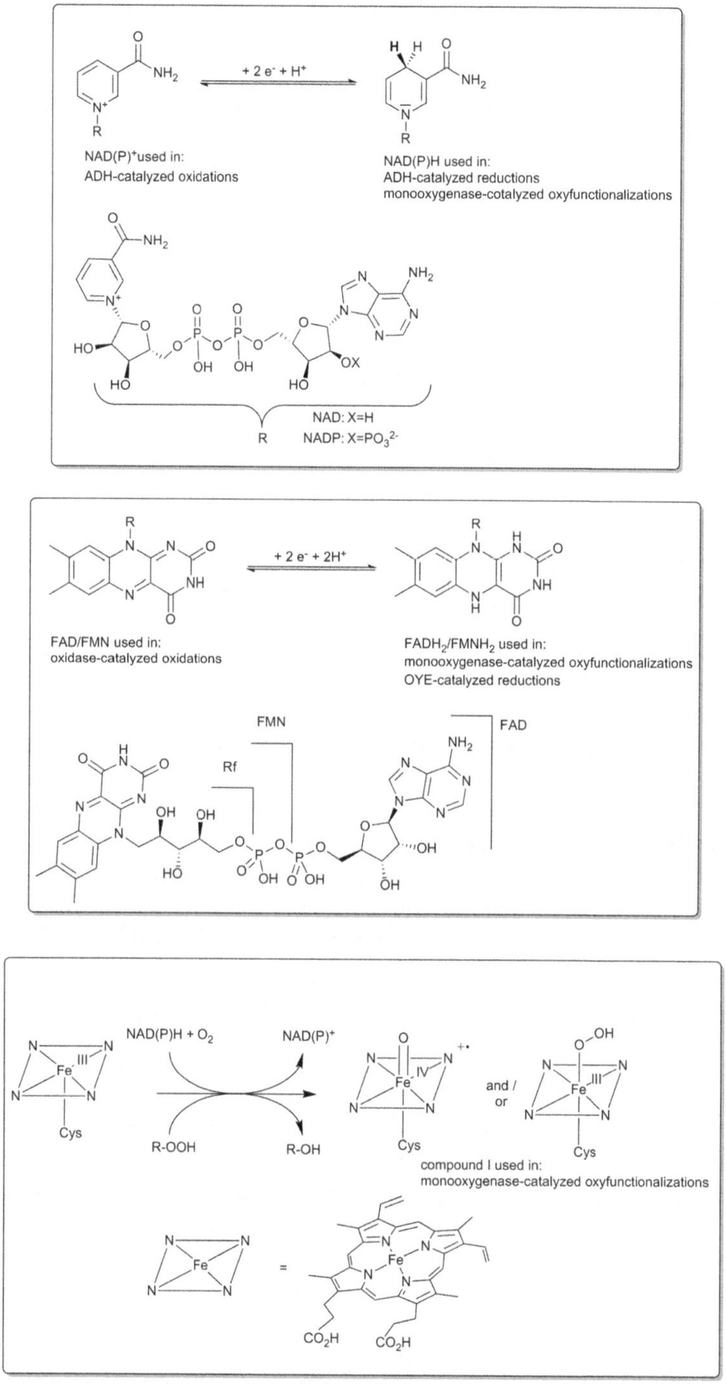

Scheme 2 Structures and basic electrochemistry of the most relevant oxidoreductase prosthetic groups and cofactors

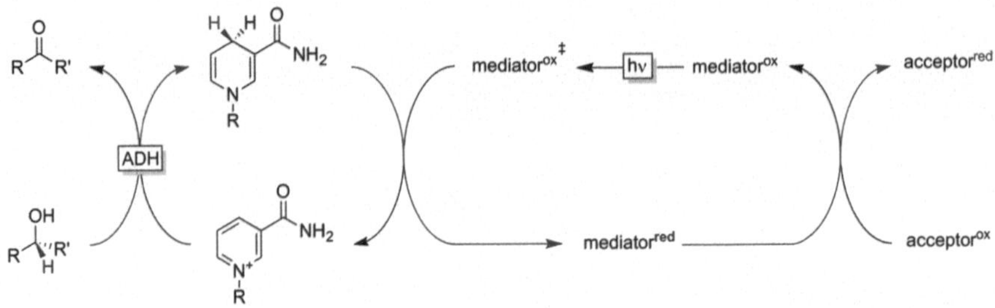

Scheme 3 General scheme for photocatalytic regeneration of oxidized nicotinamide cofactors

2 Photocatalytic Oxidative Regeneration

2.1 *NAD(P)⁺ Regeneration*

Dating back to the 1980s, photocatalytic oxidation of reduced nicotinamide cofactors to promote alcohol dehydrogenase (ADH)-catalyzed oxidations of alcohols is one of the oldest man-designed photobiocatalytic reactions [3–7]. Because of the rather negative redox potential of the NAD(P)H/NAD(P)$^+$ redox couple of −320 mV vs SCE, most NAD(P)$^+$ regeneration reactions are thermodynamically feasible (especially if O_2 serves as terminal electron acceptor). Hence the following illustrations are true examples for photocatalysis. In other words, light is used to accelerate an exergonic reaction and not as an external energy source. The most frequent photocatalytic mechanism is the photoexcitation of an oxidized mediator molecule, which in its photoexcited state reacts faster with NAD(P)H than in the corresponding ground state (Scheme 3).

Overall, a photoexcited mediator serves as hydride abstractor from NAD(P)H, yielding the reduced mediator and the (desired) oxidized nicotinamide cofactor. Whether this hydride transfer occurs concerted as a true hydride mechanism or sequentially as sequence electron transfer – deprotonation – electron transfer (ECE mechanism, see below) is only poorly understood and probably also depends on the mediator used.

To use catalytic amounts of the mediator, a sacrificial electron donor is usually applied to re-oxidize the mediator and enable the next catalytic cycle. Generally, molecular oxygen serves as terminal electron acceptor yielding either hydrogen peroxide or water as by-product. Alternatively, Handman et al reported protons as terminal electron acceptors using Pt particles as H_2-evolution catalysts (Scheme 4) [12]. A photoelectrochemical variant of this approach has also been reported [13].

A selection of frequently used redox mediators following the reaction sequence in Scheme 3 is shown in Scheme 5.

For example, methylene blue or phenazonium methyl sulphate had been reported as a very efficient NAD$^+$ regeneration catalyst under visible light illumination [14, 15]. Turnover numbers (TNs) for the nicotinamide cofactor of up to 1,125 had been reported, pointing towards a very efficient regeneration system.

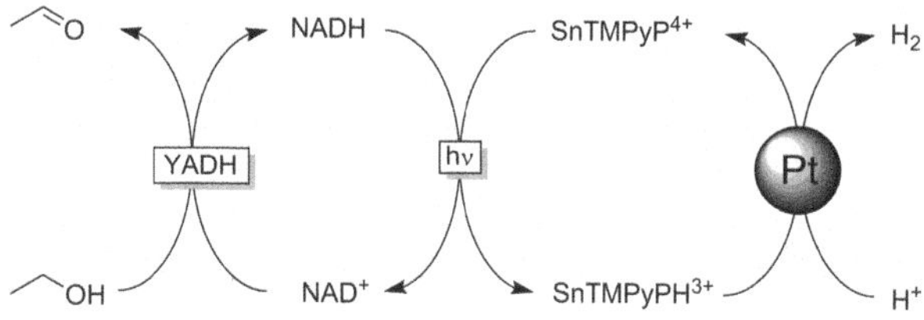

Scheme 4 Photo-chemo-enzymatic dehydrogenation of ethanol. Using the ADH from yeast (YADH), Sn-*meso*-tetrakis (*N*-methyl-4-pyridyl)porphine (SnTMPyP) as photocatalyst/mediator and Pt as H_2-evolution catalysts

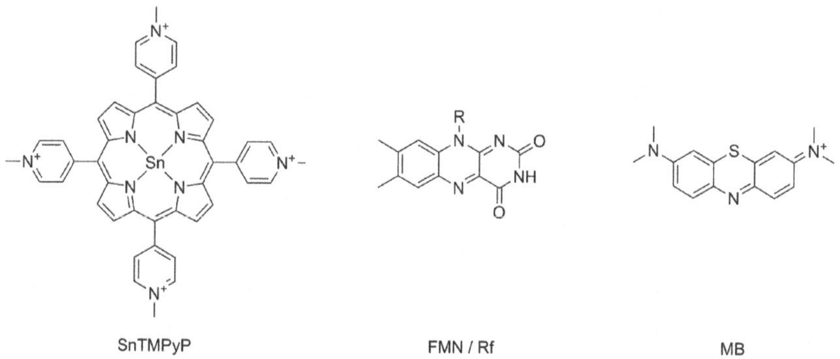

Scheme 5 Commonly used organic dyes for photoaccelerated oxidation of NAD(P)H. SnTMPyP: Sn-*meso*-tetrakis (*N*-methyl-4-pyridyl)porphine; FMN/Rf: Flavin mononucleotide/riboflavin; MB: methylene blue

Interestingly, H_2O_2 was not detected, which the authors attribute to H_2O_2 being an even more efficient reoxidant of MBH_2 than O_2. Unfortunately, this very promising system was not followed up in later studies.

More recently we reported that visible light significantly accelerated the well-known aerobic reoxidation of both NADH and NADPH by simple flavins such as FMN and riboflavin [16–19]. In the absence of an external (visible) light source the reaction kinetics are painfully slow, necessitating stoichiometric amounts of the flavin "catalyst" to enable reasonable overall rates. Simple illumination with a commercial white light bulb dramatically accelerated this process overall, enabling catalytic turnover of both the flavin catalysts and the nicotinamide cofactors [20]. Using this setup, (chiral) lactones became accessible through oxidation of diols (Scheme 6) [21–23].

A drawback of this system, however, is the formation of hydrogen peroxide, which for the sake of enzyme stability necessitated application of catalase.

Laccase mediator systems can be used for the in situ regeneration of $NAD(P)^+$ [21–26]. Also here, under certain circumstances, visible light can accelerate this process [27].

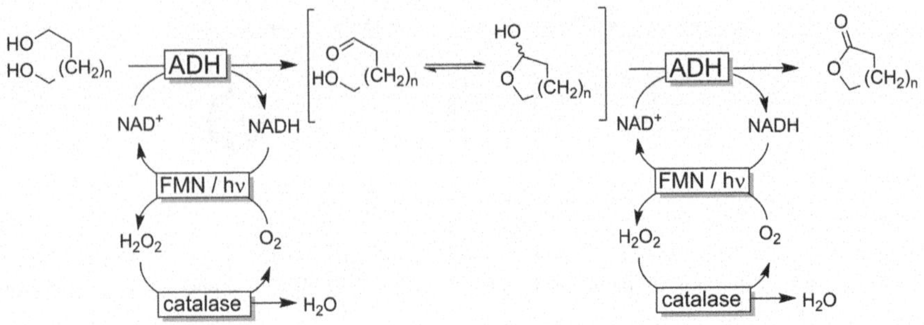

Scheme 6 Light-accelerated aerobic regeneration of NAD+ to promote ADH-catalyzed oxidative lactonization reactions

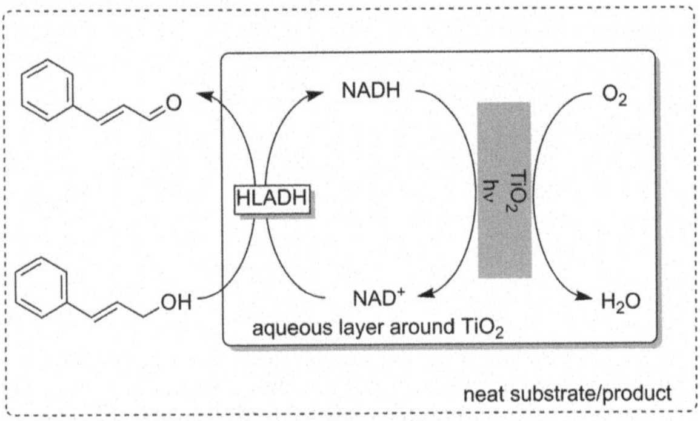

Scheme 7 Photocatalytic NAD^+ regeneration system using TiO_2 as regeneration catalyst

In addition to the aforementioned organic dyes, inorganic semiconductors have also been used as photocatalysts to accelerate the (aerobic) reoxidation of NAD(P) H. For example, TiO_2 has been reported by Tanaka and coworkers for this purpose [28]. Next to its function as photocatalyst, TiO_2 also served as carrier material to immobilize the ADH (from horse liver, HLADH) and enabled using a near-neat reaction system (Scheme 7).

This represents a very interesting approach which, however, has not been followed up much in the literature. The use of TiO_2 as light-harvesting photosensitizer is limited because of its wide optical bandgap (3.2 eV) and thus restricted application under ultraviolet light less than 387 nm.

The photocatalytic NAD(P)H oxidation systems presented so far all rely on a reductive quenching mechanism. In other words, the photoexcited catalyst is able to oxidize NAD(P)H. An alternative mechanism relies on oxidative quenching; here the oxidized mediator reacts quickly with NAD(P)H and photoexcitation acts on the reduced mediator, thereby facilitating its own reoxidation. Steckhan and coworkers pioneered this approach using a tribipyridylruthenium $[Ru(bpy)_3]^{2+}$ as photocatalyst/mediator. The Ru(III) complex swiftly oxidises NAD(P)H (in an ECE mechanism) and the resulting Ru(II) complex is photoexcited to allow for

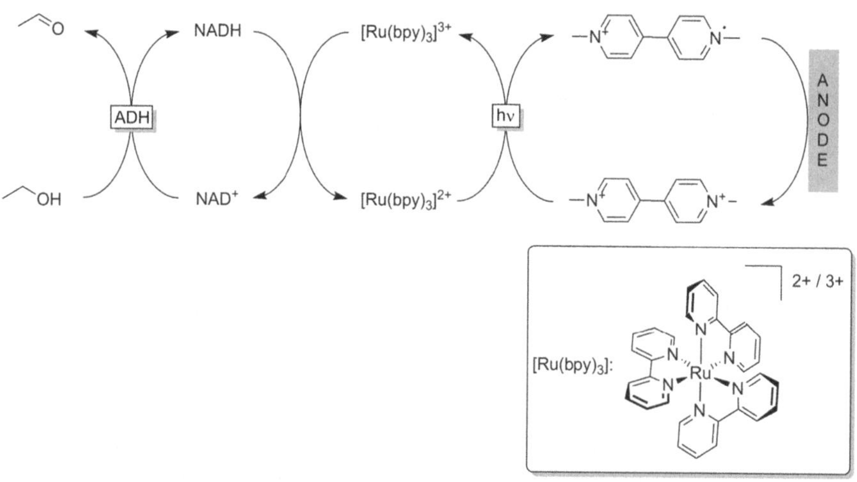

Scheme 8 Photoelectrochemical regeneration of NAD^+ to promote ADH-catalyzed oxidation reactions

reoxidation with methyl viologen. The latter is reoxidized anodically (Scheme 8) [29].

This system represents an interesting but not very practical approach (need for electrochemical and photochemical equipment, use of toxic viologenes). Furthermore, the turnover numbers (TNs) and frequencies (TFs) are too low to suggest economic feasibility.

In conclusion, a broad range of promising photochemical $NAD(P)^+$ regeneration systems has been reported. However, compared to alternative enzymatic regeneration systems, they clearly fall back in terms of popularity.

2.2 *$NAD(P)^+$ Independent Oxidative Regeneration of Oxidoreductases*

Examples of the direct oxidative regeneration are few. Gray and coworkers reported a very interesting approach to oxidize heme-Fe(III) using a covalently attached Ru-photocatalyst [30]. This approach, if further developed, may actually lead to O_2-independent P450-catalysis.

Another method of oxidative regeneration of P450 monooxygenases is to utilize the so-called hydrogen peroxide shunt pathway [31]. Here, the catalytically active oxyferryl species is formed directly from the resting state of the enzyme and H_2O_2, which circumvents the need for an expensive nicotinamide cofactor together with a regeneration system. The principal feasibility of this approach using photochemically-generated H_2O_2 has been demonstrated [32, 33]. So-called peroxygenases are (even more than P450 monooxygenases) of interest here as they utilize H_2O_2 as the natural

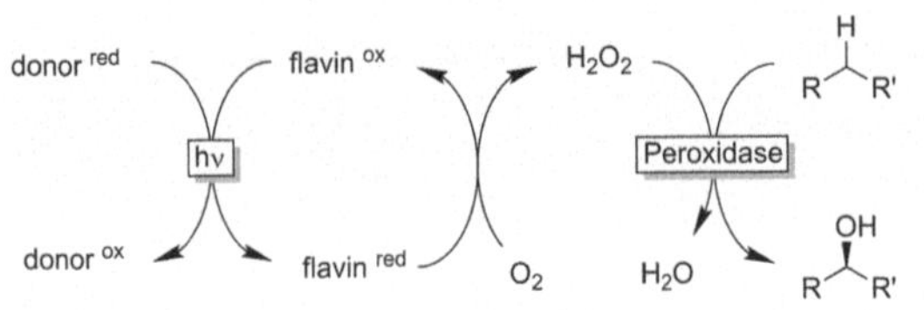

Scheme 9 Photocatalytic reduction of molecular oxygen to provide peroxidases with H_2O_2 for (stereoselective) oxyfunctionalization reactions

oxidant [34–36]. The challenge to be met here is the poor stability of heme enzymes in the presence of excess H_2O_2 leading to oxidative inactivation of the prosthetic heme group [34–36]. We have addressed this issue via photocatalytic in situ generation of H_2O_2 using a flavin photocatalyst (Scheme 9) [37–40].

Very promising catalytic performances (in terms of TFs and TNs of the catalysts applied) have been observed so far. These systems are currently under further investigation in our laboratory and we are confident that they are going to become compatible alternatives to the established enzymatic, chemical and electrochemical systems.

3 Photocatalytic Reductive Regeneration

3.1 NAD(P)H Regeneration

Reduced nicotinamide cofactors (NAD(P)H) play a central role in biocatalytic redox reactions. In nature NAD(P)H are the central reductants used for a vast range of different reduction and oxyfunctionalization reactions. Therefore it is also not surprising that in situ regeneration of NAD(P)H has also been focussed on in research for many years now. Basically, the majority of photocatalytic NAD (P)H regeneration systems can be summarized by Scheme 10.

A photosensitizer/photocatalyst is applied to liberate reducing equivalents from a sacrificial electron donor. The majority of sacrificial electron donors are low potential (high energy content) compounds, thus the electron transfer is thermodynamically feasible and the photocatalyst merely accelerates this step. Photosynthetic reactions, i.e. reaction schemes utilizing light energy to add thermodynamic driving force into an uphill electron transfer (e.g. from water mimicking natural photosynthesis), are very rare (see below). Once liberated from the sacrificial electron donor, the reducing equivalents are transferred indirectly (i.e. via a mediator and a regeneration catalyst) to $NAD(P)^+$. The need for the regeneration catalyst ($NAD(P)^+$ reduction catalyst) is because of the redox chemistry of $NAD(P)^+$, which has been investigated in detail in the past for electrochemical NAD(P)H regeneration systems [41]. In essence, a sequence of single electron transfer, protonation and single electron transfer (overall corresponding to a hydride addition

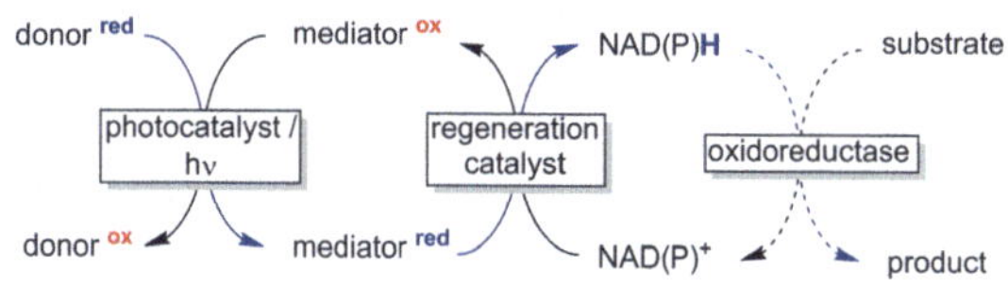

Scheme 10 General scheme of photocatalytic regeneration of reduced nicotinamide cofactors to promote NAD(P)H-dependent redox reactions

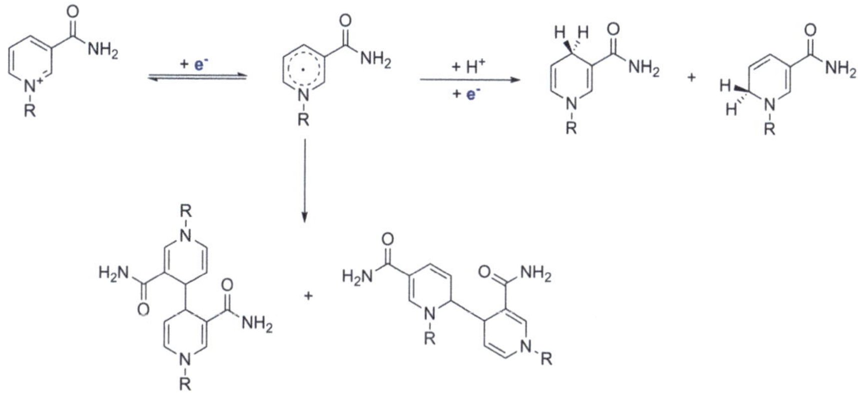

Scheme 11 Steps and products involved in the single electron reduction (either electrochemical of photochemical) of $NAD(P)^+$. The primary single electron transfer (SET) is followed by fast protonation and second SET of the intermediate radical species eventually yielding a mix of (enzymatically active) 1,4-NADH and (enzymatically inactive) 1,2- and 1-6-NADH. Radical recombination yielding NAD dimers is also observed [41]

to $NAD(P)^+$) usually results in various undesired side reactions such as dimerization and formation of enzymatically inactive 1,2- and 1,6-isomers of NAD(P)H (Scheme 11).

In the last three decades an enormous variety of photocatalytic NAD(P)H regeneration systems has been reported. A systematization can be made based on the photocatalyst, the mediator, the NAD(P)H regeneration catalyst or, of course, the sacrificial electron donors used. In the following we briefly comment on all of these aspects.

3.1.1 Photocatalysts Used for NAD(P)H Regeneration

The most popular photocatalysts are (in)organic semiconductor materials, organometallic complexes and organic dyes. Scheme 12 shows the most prominent small molecules used as photosensitizers for photochemical NAD(P)H regeneration. Especially tris bipyridine ruthenium ($[Ru(bpy)_3]^{2+}$) has been used early on, e.g. by Willner and coworkers [42–45], but finds renewed interest nowadays [46]. More recently, light-absorbing organic dyes such as eosin [47–49] and its derivatives [50], proflavine [51] and oligothiophenes [52] are in focus. Similarly,

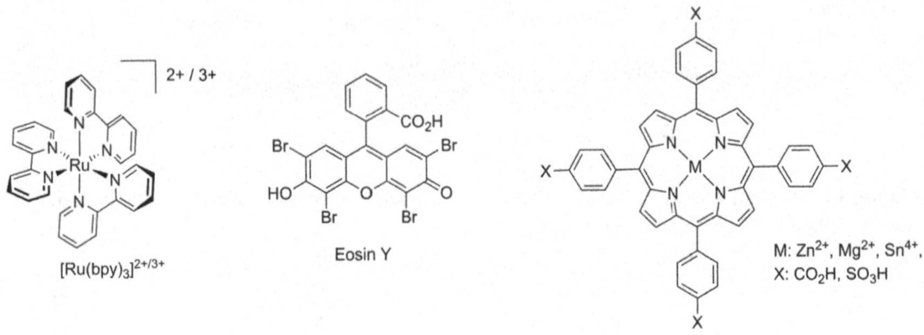

Scheme 12 Popular photocatalysts for photochemical NAD(P)H regeneration

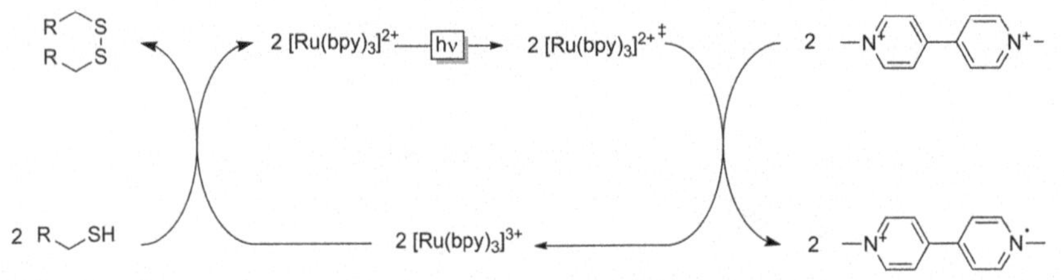

Scheme 13 General mechanism of light- and $[Ru(bpy)_3]$-promoted electron transfer from a sacrificial electron donor (e.g. thiol) to a mediator (e.g. methyl viologen). Oxidation of the sacrificial electron donor (e.g. a thiol) by the oxidized photocatalysts proceeds spontaneously. Upon absorption of a photon the resulting reduced photocatalyst is brought to an electronically excited state wherein it is able to reduce the mediator (e.g. viologen)

arylene-vinylene polymers [53], graphene derivates [54–58] and "synthetic wood" [59] are under investigation. The same is true for "bioinspired" porphyrin-based photocatalysts [60–63].

The predominant mechanism of action of the aforementioned molecular photocatalysts is oxidative quenching, i.e. upon photoactivation the reduced photocatalyst is able to reduce the intermediate electron acceptor. Scheme 13 exemplifies this with the example of the $[Ru(bpy)_3]^{2+}$-mediated electron transfer between thiols (sacrificial electron donor) and methyl viologen (mediator).

In parallel to the aforementioned molecular photocatalysts, semiconductor-based photocatalysts have also been developed. By far the most widely investigated photocatalyst is TiO_2 together with C-, P- or B-doped versions of it [64–69]. In addition, other inorganic semiconductors such as CdS [70], ZnS [71], $W_2Fe_4Ta_2O_{17}$ [72] and organic semiconductors such as graphitic carbon nitride [73–75] have been reported. The general scheme of semiconductor-based photocatalysts is depicted in Scheme 14. Interaction of the semiconductor with light leads to the promotion of an electron from the valence band into the conducting band. The "electron hole" in the conducting band is filled by oxidation of a sacrificial electron donor (predominantly TEOA) whereas the electron promoted into the valence band is eventually transferred to the mediator.

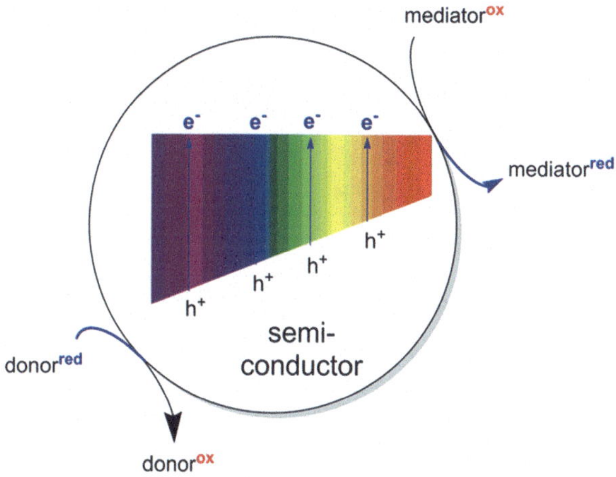

Scheme 14 Semiconductors as photocatalysts to enable the electron transfer from a sacrificial electron donor to a mediator

3.1.2 Sacrificial Electron Donors

For the catalytic reduction of $NAD(P)^+$ a stoichiometric source of electrons is necessary. For this task, two major substance classes have been established: thiols (such as mercaptoethanol) and β-aminoalcohols and β-aminoacids (such as triethanolamine or EDTA). A few other sacrificial electron donors have been summarized in Table 1.

Although thiols are mostly relevant from a 'historical' point of view, β-aminoalcohols and β-aminoacids are very popular today, possibly because of the usually high electron transfer rates observed with them. However, their use for future preparative applications has to be questioned for several reasons. First, these compounds are already "high energy compounds" and therefore the overall reaction is already thermodynamically feasible. The role of the photocatalyst and of the light energy introduced into the system therefore resides with catalysis, i.e. acceleration of a feasible reaction. Furthermore, the use of thiols or β-aminoalcohols and β-aminoacids is questionable from an atom economy point of view (i.e. generation of significant amounts of problematic by-products).

From an environmental point of view, water would be most desirable as sacrificial electron donor. Unfortunately, today, there are very few examples of this approach. Park and coworkers reported an interesting method using Co-polyoxometalates (such as $[Co_4(H_2O)_2(PW_9O_{34})_2]^{10-}$) as water oxidation catalysts together with $[Cp^*Rh(bpy)(H_2O)]^{2+}$ derivatives as NAD(P)H regeneration catalyst (see below) [46]. To make the WOC-mediated oxidation of water feasible, a $Ru(bpy)^{3+}$ catalyst was applied (Scheme 15) and its re-oxidation was enabled after visible light-absorption. Although the turnover numbers in this system are still rather low and there are further shortcoming of transition metal catalysts (discussed below), this system represents an impressive proof of concept pointing towards water-driven redox biocatalysis!

Table 1 Selection of the most common sacrificial electron donors for the photocatalytic reduction of $NAD(P)^+$

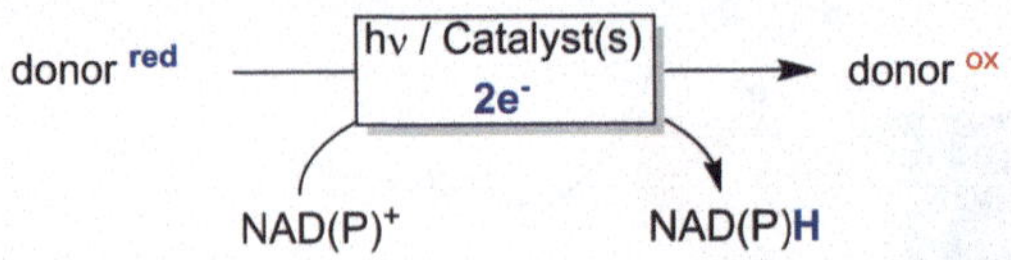

$Donor^{red}$	$Donor^{ox}$	"E-factor" ($g_{Waste} \times mol_{NAD(P)H}^{-1}$) References
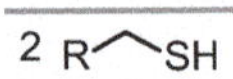	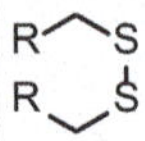	154 (Mercaptoethanol) [42–45]
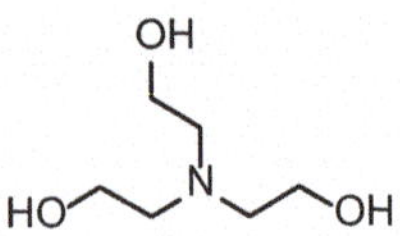	NH_3 + 6 HCHO	65.7 [47–51, 53–59, 61–63, 65, 69, 73–76]
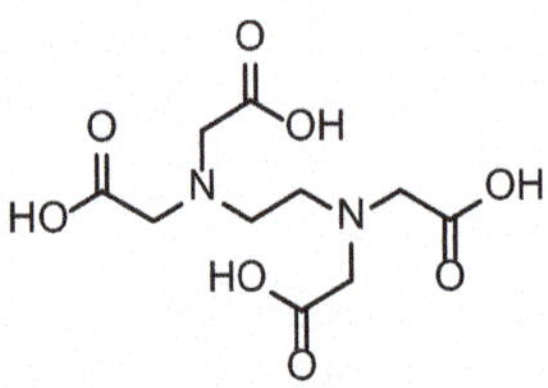	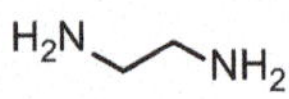 + 4 CO_2 + 4 HCHO	89 [52, 62, 72]
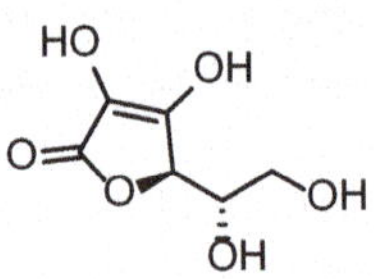	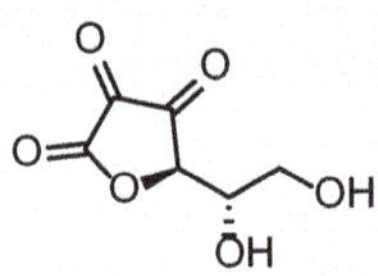	174 [60–63]
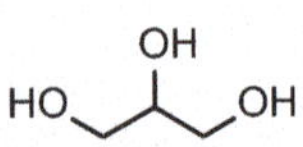	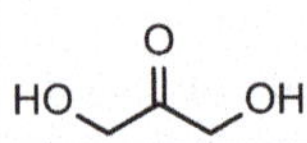	90 [71, 77]
HCO_2H	CO_2	44 [64–69]
H_2O	½ O_2	32 [46]
Cathode	–	[a] [78]

[a]No coproduct is generated. The value depends on the origin of the electrical power used

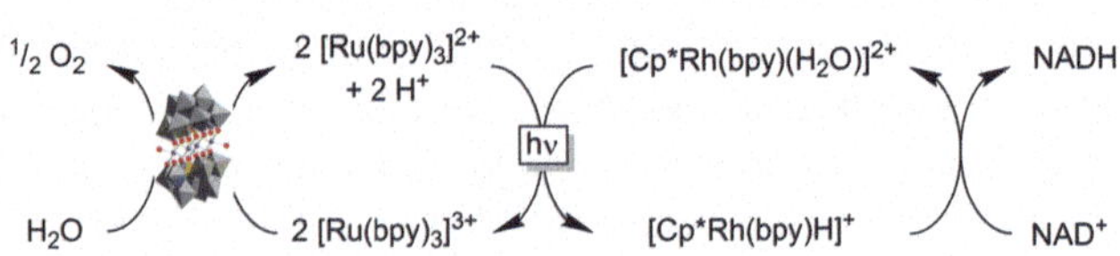

Scheme 15 Coupling of water oxidation to NADH regeneration

3.1.3 NAD(P)H Regeneration Catalysts

Once liberated from the sacrificial electron donor, the reducing equivalents have to be transferred to $NAD(P)^+$ to attain the desired in situ regeneration of the reduced cofactors (NAD(P)H). As mentioned above, this mostly cannot be achieved by direct electron transfer from the reduced photocatalyst to $NAD(P)^+$ as this electron transfer mostly comprises single electron transfer (SET) steps with the mechanistic challenges associated to it (Scheme 11). Therefore, a relay system transforming two SET steps and a protonation step into a single hydride step is necessary for efficient NAD(P)H regeneration. According to Steckhan such a NAD(P)H regeneration catalyst has to fulfil a range of requirements [79]. First, it has to act as hydride donor instead of mediating single electron transfer steps; this is to avoid NAD (P) radical formation and the undesired side reactions resulting thereof (Scheme 11). Second, the catalytically active form has to be formed at redox potentials less negative than the first $NAD(P)^+$ reduction potential (to avoid direct reduction of $NAD(P)^+$ by the source of reducing equivalents). Third, a successful NAD(P)H regeneration catalyst must selectively form the enzymatically active 1,4-NADH isomer only. Finally, the NAD(P)H regeneration catalysts should not interfere with the other reactants in the system.

Today, principally two different NAD(P)H regeneration systems are available: (1) ferredoxin-$NADP^+$ reductase (FDR) and (2) pentamethylcyclopentadienyl Rh complexes ($[Cp^*Rh(bpy)(H_2O)]^{2+}$).

Ferredoxin-$NADP^+$ Reductase (FDR)

FDR (EC 1.18.1.2) is a flavoprotein whose natural role appears to be to mediate the electron transfer between ferredoxin (FD) and NADP. A catalytic flavin prosthetic group is responsible for the conversion of the two SET steps to a hydride step, which is enabled by the flexible redox chemistry of flavins (Scheme 16).

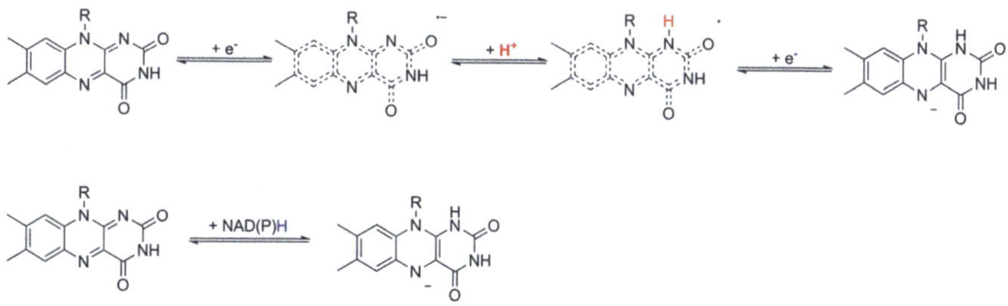

Scheme 16 Simplified flavin redox chemistry. *Top*: Flavin reduction through two SET steps (via an intermediate, stabilized semiquinone radical). *Bottom*: Flavin reduction through (e.g. NAD(P) H-mediated) one-step hydride transfer

This reaction is reversible, which is why FDR mediates both the reduction of $NADP^+$ by two reduced FDs and the oxidation of NADPH by two oxidized FDs. FDR is highly specific with respect to the nicotinamide cofactor used (accepting only the phosphorylated form) but exhibits a very relaxed substrate scope with respect to the one electron donor/acceptor. Next to ferredoxin, simple metals complexes and salts as well as a range of quinones are also converted. An exception to this is the previously mentioned photocatalyst $[Ru(bpy)_3]^{2+}$. Therefore, all photocatalytic systems using FDR as NADPH regeneration system utilize additional mediators (facilitating the electron transfer from the reduced photocatalysts to FDR) [42–45, 52, 60–69]. Of these, methyl viologen, as pioneered by Willner and coworkers, is by far the most popular and efficient [42–45, 52, 60–69]. Very promising catalytic turnovers of the single catalysts have been reported as early as the 1980s, reaching several thousands for $[Ru(bpy)_3]^{2+}$ and hundred thousands for FDR. The turnover numbers for NADP and the viologen mediator are less promising, ranging from a few dozen to hundreds. These lower numbers together with the high toxicity of viologenes and the limitation to FDR to NADPH-regeneration possibly explains why at present this approach is not the dominating one.

$[Cp^*Rh(bpy)(H_2O)]^{2+}$

The organometallic compound $[Cp^*Rh(bpy)(H_2O)]^{2+}$ was reported as early as the 1980s by Steckhan and coworkers to be an efficient catalysts for indirect electrochemical regeneration of NAD(P)H [80–82]. Several features make $[Cp^*Rh(bpy)(H_2O)]^{2+}$ and its derivates interesting catalysts for the photochemical regeneration of NAD(P)H. (1) $[Cp^*Rh(bpy)(H_2O)]^{2+}$ is capable of both single electron and of hydride transfer reactions. Hence, it can accept two successive electrons (and one proton) from an electrochemical or a photochemical cathode forming a hydrido species ($[Cp^*Rh(bpy)H]^+$) serving as hydride reductant for $NAD(P)^+$, thereby avoiding NAD(P) radicals and the undesired side reactions associated with it (Scheme 11). (2) The active hydrido species coordinates to the carbonyl group of $NAD(P)^+$ and thereby is positioned close to the C4-atom, resulting in highly regioselective hydride transfer and minimizing the undesired formation of enzymatically inactive 1,2- and 1,6-NAD(P)H isomers [83]. Therefore, it is hardly surprising that $[Cp^*Rh(bpy)(H_2O)]^{2+}$ is frequently used as NAD(P)H regeneration catalyst [46, 48–51, 53–58, 61, 62, 65–67, 69–73, 76, 78, 84, 85].

Despite its great success in non-enzymatic NAD(P)H regeneration, it should be mentioned here that $[Cp^*Rh(bpy)(H_2O)]^{2+}$ exhibits some significant drawbacks which in the long term may severely limit its practical usefulness. First, as with many transition metals, the Rh central atom is rather expensive (and future projections of the Rh prices point towards even higher prices). At the same time, turnover numbers reported for this catalyst (so far) range between 10 and 1,000. As a consequence, the cost contribution of this catalyst alone to the final product is very significant. Certainly, cheaper catalysts (e.g. based on abundant metals) with improved catalytic performance would give the field a fresh impetus! Another

Table 2 Selection of photochemical NAD(P)H regeneration systems

Photosensitizer	Mediator	Electron donor	NAD(P)H regeneration rate (mM h^{-1})	TN (NAD (P))	References
Eosin Y	$[Cp^*Rh(bpy)(H_2O)]^{2+}$	TEOA	1.42	23	[48]
ZnTPPS	$[Cp^*Rh(bpy)(H_2O)]^{2+}$	TEOA	0.23	n.d.	[61]
$Ru(bpy)_3{}^{2+}$	FDR	Mercaptoethanol	n.d.	15	[44]
CCGCMAQSP	$[Cp^*Rh(bpy)(H_2O)]^{2+}$	TEOA	0.091	89	[54]
Graphene-BODIPY	$[Cp^*Rh(bpy)(H_2O)]^{2+}$	TEOA	0.11	116	[57]
Proflavin	$[Cp^*Rh(bpy)(H_2O)]^{2+}$	TEOA	1.28	33	[51]
Synthetic wood	$[Cp^*Rh(bpy)(H_2O)]^{2+}$	TEOA	0.28	4.5	[59]
Carbon nitride	$[Cp^*Rh(bpy)(H_2O)]^{2+}$	TEOA	2	5	[74]
H-SiNWs	$[Cp^*Rh(bpy)(H_2O)]^{2+}$	TEOA	0.4	1.4	[76]
CdS/TiO_2	$[Cp^*Rh(bpy)(H_2O)]^{2+}$	TEOA	0.25	3.6	[69]
SnC	$[Cp^*Rh(bpy)(H_2O)]^{2+}$	EDTA	0.056	1.0	[62]

TEOA triethanolamine, *CCGCMAQSP* chemically converted graphene coupled with multianthraquinone-substituted porhyrin, *FDR* ferredoxin-NADP$^+$ reductase; BODIPY = (1-picolylamine-2-aminophenyl-3-oxy-phenyl-4,40-difluoro-1,3,5,7-tetramethyl-2,6-diethyl-4-bora-3a,4a-diaza-s-indacene-triazine), H-SiNWs = hydrogen-terminated silicon nanowires, SnC = tin(IV)-meso-tetrakis(*N*-methylpyridinium)-chlorin

issue related to $[Cp^*Rh(bpy)(H_2O)]^{2+}$ (and other transition metals) is the frequently observed mutual inactivation of the metal catalyst and the biocatalyst, which needs further attention [86, 87]. In this respect, new catalysts such as cobaloximes may become interesting in the future [47].

Finally, Table 2 gives a representative overview of the performance of some photobiocatalytic systems involving NAD(P)H regeneration. The TN reported for the nicotinamide and for the photo- and NAD(P)H-regeneration catalysts are still orders of magnitude away from economic feasibility.

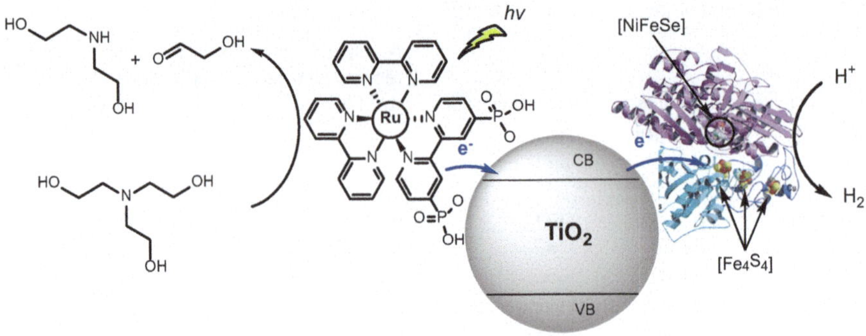

Scheme 17 Schematic representation of photocatalytic hydrogen generation with a [NiFeSe]-hydrogenase and Ru dye-sensitized TiO_2

4 Direct Regeneration of Oxidoreductases

NAD(P)H is the most important redox mediator in all living systems. However, many oxidoreductases do not directly rely on NAD(P)H as reductant. For example, monooxygenases utilize NAD(P)H to generate reduced species (e.g. reduced heme or flavin species) that activate molecular oxygen for the actual oxygenation reaction. Hence, NAD(P)H "only" serves as reductant and is not directly involved in the catalytic mechanism. Substitution of NAD(P)H by artificial reduction catalysts appears to be a straightforward method to simplify the catalytic mechanism substantially.

Two catalyst systems, (1) flavins and (2) Ru complexes, are discussed in somewhat more detail below. Besides these two examples, Park and coworkers recently reported on eosin Y to regenerate P450 BM3 directly [88]. In another example, Reisner and coworkers developed a photocatalytic hydrogen evolution system in which a [NiFeSe]-hydrogenase was attached directly to Ru dye-sensitized TiO_2 nanoparticles (Scheme 17) [89–91]. Other photosensitizers such as polymeric carbon nitride, eosin Y and cyanobacterial photosystem II were also utilized to regenerate the hydrogenase directly [92–97].

Cheruzel and coworkers have extensively studied various Ru (II) photosensitizers to reduce directly heme groups in P450 monooxygenases for catalysis (Scheme 18) [98–100]. Variations of the Ru(II)-photosensitizers described above were covalently linked to the heme-dependent monooxygenase from *Bacillus megaterium* (P450 BM3), which enabled efficient direct electron transfer from (in situ photogenerated Ru(I)) to the heme iron.

Alternatively, flavins (such as riboflavin, flavin mononucleotide, FMN or flavin adenine dinucleotide, FAD) can also be used as photosensitizer/reductant for enzymes. Flavins are a class of biological cofactors which act as a redox centre for many oxidoreductases. They themselves are photoexcitable under visible light and thus can act as photosensitizers for the reductive regeneration of flavin-containing enzymes for catalysis. This was first demonstrated by us, using a

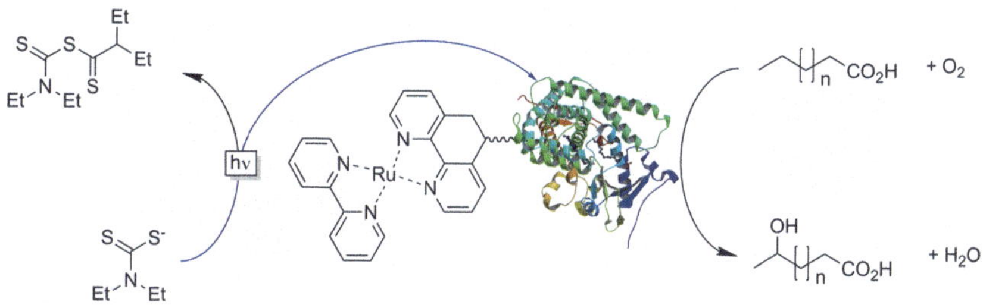

Scheme 18 Direct photochemical reduction of the heme group of P450 BM3 to achieve reductive activation of molecular oxygen for catalysis

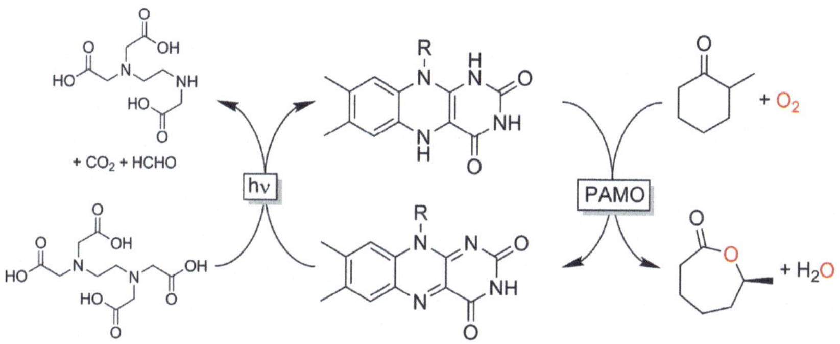

Scheme 19 Using FAD as photosensitizer to couple EDTA oxidation to a monooxygenase reaction

flavin-dependent monooxygenase (phenylacetone monooxygenase, PAMO) [101]. Upon photoexcitation, flavins are capable of utilizing simple reductants, e.g. mentioned in Table 1, and transferring the electrons to the enzymes' active sites (Scheme 19) [102].

Even though the principal feasibility of this setup could be demonstrated with promising conversions and enantioselectivities, the performance of the artificial system fell back significantly behind the performance of the natural cycle using NADPH. Oxidative decoupling of the regeneration reaction from the enzymatic reaction produced by the rapid reaction of reduced flavin species with molecular oxygen probably accounts for the poor performance [32, 33, 103, 104]. In fact, less than 10% of the electrons provided by the sacrificial electron donor (EDTA) were used productively, i.e. for the enzymatic Baeyer–Villiger oxidation. The majority of reducing equivalents was channelled into direct oxygen reduction yielding H_2O_2. A possible solution to this *Oxygen Dilemma* may be to use deazaflavins [105]. Alternatively, the *Oxygen Dilemma* can be used productively by using the H_2O_2 generated to promote peroxygenase-catalyzed oxyfunctionalization reactions (see above).

A third alternative is simply to circumvent the *Oxygen Dilemma* by utilizing O_2-independent enzymes such as Old Yellow Enzymes (OYEs) for enantioselective

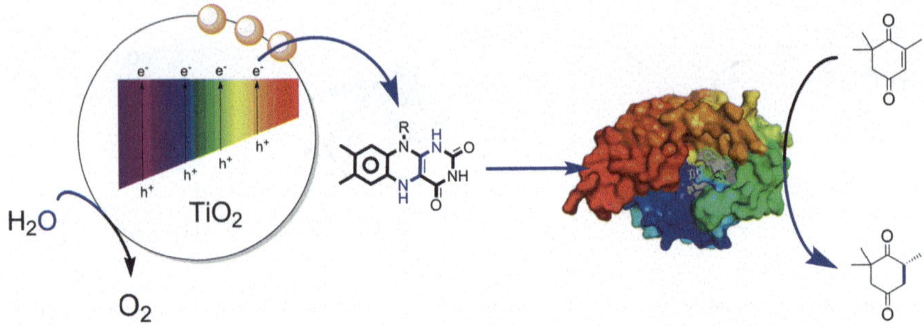

Scheme 20 Photochemical water oxidation used to promote biocatalytic reduction reactions

reduction of conjugated C=C bonds [106–108]. Members of this enzyme class contain a catalytic flavin (FMN) in their active site, which in its reduced form performs a Michael-type hydride addition to conjugated C=C bonds. Various studies have demonstrated that the native nicotinamide cofactor may be substituted by other reductants such as viologenes [109], "smart cosubstrates" [110, 111], chemically reduced flavins [112] or synthetic nicotinamide analogues [113].

The reduction of OYEs using flavins as photocatalysts and electron mediators was established using EDTA as sacrificial electron donor [101, 114]. Indeed, it could be demonstrated that the electron transfer yield (i.e. the percentage of electrons used productively) in the absence of molecular oxygen was close to the theoretical value. Furthermore, the enzyme properties (such as enantioselectivity or enzyme activity) were apparently not impaired by the "unnatural" reaction conditions. One drawback from an environmental point of view, however, was the nature of the by-products formed (see Table 1). However, we demonstrated very recently that productive coupling of the OYE-catalyzed reduction reaction to a photochemically operated water oxidation catalyst is feasible (Scheme 20).

5 Conclusions

Photobiocatalysis is a dynamic and rapidly evolving area of research. As outlined in this chapter, many different approaches are currently being explored by various research groups.

The scope of photobiocatalysis lies in more robust and simplified regeneration schemes, especially if shortened electron transport chains are enabled by direct regeneration of the oxidoreductases' active sites. Today, however, the traditional enzymatic regeneration systems still outperform their photochemical counterparts in terms of turnover numbers of the catalytic components used. We need to intensify our research efforts to improve the efficiency of the reaction systems. A particular focus here should concentrate on (1) improving the turnover numbers of the catalysts used (only if the catalysts are used efficiently, i.e. if their cost contribution to the final product is sufficiently low, can "real life" application of the systems developed occur) and (2) increasing the reagent payload (today, most reports deal with low substrate concentrations around 10 mM, which clearly is orders of magnitude away from practical feasibility). The time for proof-of-concept studies is slowly running out; now is the time to make the systems truly practical!

Acknowledgements Financial support by the European Union (ERC consolidator grant No 648026) is gratefully acknowledged.

References

1. Hollmann F, Arends IWCE, Holtmann D (2011) Green Chem 13:2285
2. Hollmann F, Arends IWCE, Buehler K, Schallmey A, Buhler B (2011) Green Chem 13:226
3. Chenault H, Whitesides G (1987) Appl Biochem Biotechnol 14:147
4. Whitesides GM, Wong CH (1985) Angew Chem Int Ed Engl 24:617. doi:10.1002/anie.198506173
5. Kara S, Schrittwieser JH, Hollmann F, Ansorge-Schumacher MB (2014) Appl Microbiol Biotechnol 98:1517. doi:10.1007/s00253-013-5441-5
6. Goldberg K, Schroer K, Lutz S, Liese A (2007) Appl Microbiol Biotechnol 76:237. doi:10.1007/s00253-007-1002-0
7. Weckbecker A, Groger H, Hummel W (2010) Biosystems engineering I: creating superior biocatalysts. Springer, Berlin, p 195
8. Maciá-Agulló JA, Corma A, Garcia H (2015) Chem Eur J. 10.1002/chem.201406437
9. Hollmann F, Arends Isabel WCE, Buehler K (2010) ChemCatChem 2:762
10. Kochius S, Magnusson A, Hollmann F, Schrader J, Holtmann D (2012) Appl Microbiol Biotechnol 93:2251. doi:10.1007/s00253-012-3900-z
11. Quinto T, Köhler V, Ward T (2013) Top Catal 1. 10.1007/s11244-013-0187-y
12. Handman J, Harriman A, Porter G (1984) Nature 307:534
13. Hambourger M, Brune A, Gust D, Moore AL, Moore TA (2005) Photochem Photobiol 81:1015. doi:10.1562/2005-05-15-rc-528r.1
14. Julliard M, Le Petit J, Ritz P (1986) Biotechnol Bioeng 28:1774
15. Julliard M, Lepetit J (1982) Photochem Photobiol 36:283. doi:10.1111/j.1751-1097.1982.tb04376.x
16. Bridges AJ, Raman PS, Ng GSY, Jones JB (1984) J Am Chem Soc 106:1461
17. Jakovac IJ, Goodbrand HB, Lok KP, Jones JB (1982) J Am Chem Soc 104:4659
18. Jones JB, Taylor KE (1976) Can J Chem 54:2969
19. Jones JB, Taylor KE (1973) J Chem Soc Chem Commun 205
20. Gargiulo S, Arends IWCE, Hollmann F (2011) ChemCatChem 3:338. doi:10.1002/cctc.201000317
21. Könst P, Kara S, Kochius S, Holtmann D, Arends IWCE, Ludwig R, Hollmann F (2013) ChemCatChem 5:3027. doi:10.1002/cctc.201300205
22. Kara S, Spickermann D, Schrittwieser JH, Weckbecker A, Leggewie C, Arends IWCE, Hollmann F (2013) ACS Catal 3:2436. doi:10.1021/cs400535c
23. Kara S, Spickermann D, Schrittwieser JH, Leggewie C, Van Berkel WJH, Arends IWCE, Hollmann F (2013) Green Chem 15:330
24. Pham NH, Hollmann F, Kracher D, Preims M, Haltrich D, Ludwig R (2015) J Mol Catal B Enzym 120:38. doi:10.1016/j.molcatb.2015.06.011
25. Aksu S, Arends IWCE, Hollmann F (2009) Adv Synth Catal 351:1211

26. Ferrandi EE, Monti D, Patel I, Kittl R, Haltrich D, Riva S, Ludwig R (2012) Adv Synth Catal 354:2821. doi:10.1002/adsc.201200429
27. Kochius S, Ni Y, Kara S, Gargiulo S, Schrader J, Holtmann D, Hollmann F (2014) ChemPlusChem 79:1554. doi:10.1002/cplu.201402152
28. Kawamoto T, Aoki A, Sonomoto K, Tanaka A (1989) J Ferment Bioeng 67:361
29. Ruppert R, Steckhan E (1989) J Chem Soc Perkin Trans 2:811
30. Immoos CE, Di Bilio AJ, Cohen MS, Van der Veer W, Gray HB, Farmer PJ (2004) Inorg Chem 43:3593. doi:10.1021/ic049741h
31. Joo H, Lin ZL, Arnold FH (1999) Nature 399:670
32. Girhard M, Kunigk E, Tihovsky S, Shumyantseva VV, Urlacher VB (2013) Biotechnol Appl Biochem 60:111
33. Zachos I, Gassmeyer S, Bauer D, Sieber V, Hollmann F, Kourist R (2015) Chem Commun 51:1918. doi:10.1039/c4cc07276f
34. Hofrichter M, Ullrich R (2014) Curr Opin Chem Biol 19:116. doi:10.1016/j.cbpa.2014.01.015
35. Bormann S, Gomez Baraibar A, Ni Y, Holtmann D, Hollmann F (2015) Catal Sci Technol 5:2038. doi:10.1039/C4CY01477D
36. Holtmann D, Fraaije MW, Opperman DJ, Arends IWCE, Hollmann F (2014) Chem Commun 50:13180. doi:10.1039/c3cc49747j
37. Churakova E, Arends IWCE, Hollmann F (2013) ChemCatChem 5:565. doi:10.1002/cctc.201200490
38. Churakova E, Kluge M, Ullrich R, Arends I, Hofrichter M, Hollmann F (2011) Angew Chem Int Ed 50:10716. doi:10.1002/anie.201105308
39. Perez DI, Mifsud Grau M, Arends IWCE, Hollmann F (2009) Chem Commun 6848
40. Sabuzi F, Churakova E, Galloni P, Wever R, Hollmann F, Floris B, Conte V (2015) Eur J Inorg Chem 3519. 10.1002/ejic.201500086
41. Hollmann F, Schmid A (2004) Biocatal Biotransformation 22:63. doi:10.1080/10242420410001692778
42. Mandler D, Willner I (1986) J Chem Soc Chem Commun 851. doi:10.1039/c39860000851
43. Willner I, Mandler D, Riklin A (1986) J Chem Soc Chem Commun 1022. 10.1039/c39860001022
44. Mandler D, Willner I (1986) J Chem Soc Perkin Trans 2:805. doi:10.1039/p29860000805
45. Mandler D, Willner I (1988) J Chem Soc Perkin Trans 2:997. doi:10.1039/p29880000997
46. Ryu J, Nam DH, Lee SH, Park CB (2014) Chem Eur J. doi:10.1002/chem.201403301
47. Kim JA, Kim S, Lee J, Baeg JO, Kim J (2012) Inorg Chem 51:8057. doi:10.1021/ic300185n
48. Lee SH, Nam DH, Kim JH, Baeg J-O, Park CB (2009) Chembiochem 10:1621
49. Lee SH, Kwon Y-C, Kim D-M, Park CB (2013) Biotechnol Bioeng 110:383. doi:10.1002/bit.24729
50. Lee SH, Nam DH, Park CB (2009) Adv Synth Catal 351:2589. doi:10.1002/adsc.200900547
51. Nam DH, Park CB (2012) Chembiochem 13:1278. doi:10.1002/cbic.201200115
52. Kim Y, Ikebukuro K, Muguruma H, Karube I (1998) J Biotechnol 59:213. doi:10.1016/s0168-1656(97)00155-7
53. Oppelt KT, Gasiorowski J, Egbe DAM, Kollender JP, Himmelsbach M, Hassel AW, Sariciftci NS, Knor G (2014) J Am Chem Soc 136:12721. doi:10.1021/ja506060u
54. Yadav RK, Baeg JO, Oh GH, Park NJ, Kong KJ, Kim J, Hwang DW, Biswas SK (2012) J Am Chem Soc 134:11455. doi:10.1021/ja3009902
55. Choudhury S, Baeg JO, Park NJ, Yadav RK (2012) Angew Chem Int Ed 51:11624. doi:10.1002/anie.201206019
56. Yadav RK, Oh GH, Park NJ, Kumar A, Kong KJ, Baeg JO (2014) J Am Chem Soc 136:16728. doi:10.1021/ja509650r
57. Yadav RK, Baeg JO, Kumar A, Kong KJ, Oh GH, Park NJ (2014) J Mater Chem A 2:5068. doi:10.1039/c3ta14442a
58. Choudhury S, Baeg JO, Park NJ, Yadav RK (2014) Green Chem 16:4389–4400. doi:10.1039/c4gc00885e

59. Lee M, Kim JH, Lee SH, Lee SH, Park CB (2011) ChemSusChem 5:581. doi:10.1002/cssc.201100074
60. Asada H, Itoh T, Kodera Y, Matsushima A, Hiroto M, Nishimura H, Inada Y (2001) Biotechnol Bioeng 76:86. doi:10.1002/bit.1029
61. Kim JH, Lee SH, Lee JS, Lee M, Park CB (2011) Chem Commun 47:10227
62. Oppelt KT, Woss E, Stiftinger M, Schofberger W, Buchberger W, Knor G (2013) Inorg Chem 52:11910. doi:10.1021/ic401611v
63. Kim JH, Lee M, Lee JS, Park CB (2012) Angew Chem 124:532. doi:10.1002/ange.201103244
64. Goren Z, Lapidot N, Willner I (1988) J Mol Catal 47:21. doi:10.1016/0304-5102(88)85069-7
65. Lee SH, Ryu J, Nam DH, Park CB (2011) Chem Commun 47:4643
66. Jiang ZY, Lu CQ, Wu H (2005) Ind Eng Chem Res 44:4165. doi:10.1021/ie049155w
67. Shi Q, Yang D, Jiang Z, Li J (2006) J Mol Catal B Enzym 43:44
68. Chen D, Yang D, Wang Q, Jiang ZY (2006) Ind Eng Chem Res 45:4110. doi:10.1021/ie0600902
69. Ryu J, Lee SH, Nam DH, Park CB (2011) Adv Mater 23:1883. doi:10.1002/adma.201004576
70. Nam DH, Lee SH, Park CB (2010) Small 6:922. doi:10.1002/smll.201000077
71. Dibenedetto A, Stufano P, Macyk W, Baran T, Fragale C, Costa M, Aresta M (2012) ChemSusChem 5:373. doi:10.1002/cssc.201100484
72. Park CB, Lee SH, Subramanian E, Kale BB, Lee SM, Baeg JO (2008) Chem Commun 5423. 10.1039/b808256a
73. Liu J, Cazelles R, Chen ZP, Zhou H, Galarneau A, Antonietti M (2014) Phys Chem Chem Phys 16:14699. doi:10.1039/c4cp01348d
74. Huang JH, Antonietti M, Liu J (2014) J Mater Chem A 2:7686. doi:10.1039/c4ta00793j
75. Liu J, Antonietti M (2013) Energy Environ Sci 6:1486. doi:10.1039/c3ee40696b
76. Lee HY, Ryu J, Kim JH, Lee SH, Park CB (2012) ChemSusChem 5:2129. doi:10.1002/cssc.201200251
77. Aresta M, Dibenedetto A, Baran T, Angelini A, Labuz P, Macyk W (2014) Beilstein J Org Chem 10:2556. doi:10.3762/bjoc.10.267
78. Lee SH, Ryu GM, Nam DH, Kim JH, Park CB (2014) ChemSusChem 7:3007. doi:10.1002/cssc.201402469
79. Steckhan E (1994) Electrochemistry V. Springer, Berlin, pp 33–83
80. Steckhan E, Herrmann S, Ruppert R, Dietz E, Frede M, Spika E (1991) Organometallics 10:1568
81. Steckhan E, Herrmann S, Ruppert R, Thommes J, Wandrey C (1990) Angew Chem Int Ed Engl 29:388
82. Ruppert R, Herrmann S, Steckhan E (1988) J Chem Soc Chem Commun 1150
83. Lo HC, Buriez O, Kerr JB, Fish RH (1999) Angew Chem Int Ed 38:1429
84. Lee SH, Lee HJ, Won K, Park CB (2012) Chem Eur J 18:5490. doi:10.1002/chem.201200281
85. Lee HJ, Lee SH, Park CB, Won K (2011) Chem Commun 47:12538
86. Poizat M, Arends IWCE, Hollmann F (2010) J Mol Catal B Enzym 63:149
87. Hildebrand F, Lütz S (2009) Chem Eur J 15:4998
88. Park JH, Lee SH, Cha GS, Choi DS, Nam DH, Lee JH, Lee JK, Yun CH, Jeong KJ, Park CB (2014) Angew Chem Int Ed Engl. 10.1002/anie.201410059
89. Reisner E, Fontecilla-Camps JC, Armstrong FA (2009) Chem Commun 5:550
90. Reisner E, Powell DJ, Cavazza C, Fontecilla-Camps JC, Armstrong F (2009) J Am Chem Soc 131:18457
91. Woolerton TW, Sheard S, Reisner E, Pierce E, Ragsdale SW, Armstrong FA (2010) J Am Chem Soc 132:2132
92. Kato M, Cardona T, Rutherford AW, Reisner E (2013) J Am Chem Soc 135:10610
93. Sakai T, Mersch D, Reisner E (2013) Angew Chem Int Ed 52:12313
94. Caputo CA, Gross MA, Lau VW, Cavazza C, Lotsch BV, Reisner E (2014) Angew Chem Int Ed 53:11538

95. Mersch D, Lee C-Y, Zhang JZ, Brinkert K, Fontecilla-Camps JC, Rutherford AW, Reisner E (2015) J Am Chem Soc 137:8541
96. Caputo CA, Wang L, Beranek R, Reisner E (2015) Chem Sci 6:5690
97. Martindale BCM, Hutton GAM, Caputo CA, Reisner E (2015) J Am Chem Soc 137:6018
98. Tran NH, Nguyen D, Dwaraknath S, Mahadevan S, Chavez G, Nguyen A, Dao T, Mullen S, Nguyen TA, Cheruzel LE (2013) J Am Chem Soc 135:14484. doi:10.1021/ja409337v
99. Tran N-H, Huynh N, Chavez G, Nguyen A, Dwaraknath S, Nguyen T-A, Nguyen M, Cheruzel L (2012) J Inorg Biochem 115:50. doi:10.1016/j.jinorgbio.2012.05.012
100. Tran N-H, Huynh N, Bui T, Nguyen Y, Huynh P, Cooper ME, Cheruzel LE (2011) Chem Commun 47:11936
101. Hollmann F, Taglieber A, Schulz F, Reetz MT (2007) Angew Chem Int Ed 46:2903
102. Frisell WR, Chung CW, Mackenzie CG (1959) J Biol Chem 234:1297
103. Massey V (1994) J Biol Chem 269:22459
104. Taglieber A, Schulz F, Hollmann F, Rusek M, Reetz MT (2008) Chembiochem 9:565
105. Zilly FE, Taglieber A, Schulz F, Hollmann F, Reetz MT (2009) Chem Commun 7152
106. Toogood HS, Scrutton NS (2014) Curr Opin Chem Biol 19:107. doi:10.1016/j.cbpa.2014.01.019
107. Toogood HS, Knaus T, Scrutton NS (2013) ChemCatChem 6:951. doi:10.1002/cctc.201300911
108. Toogood HS, Gardiner JM, Scrutton NS (2010) ChemCatChem 2:892
109. Burai TN, Panay AJ, Zhu H, Lian T, Lutz S (2012) ACS Catal 2:667. doi:10.1021/cs300085h
110. Winkler CK, Clay D, van Heerden E, Faber K (2013) Biotechnol Bioeng 110:3085. doi:10.1002/bit.24981
111. Stueckler C, Reiter TC, Baudendistel N, Faber K (2010) Tetrahedron 66:663
112. Bernard J, Van Heerden E, Arends IWCE, Opperman DJ, Hollmann F (2012) ChemCatChem 4:196
113. Paul CE, Gargiulo S, Opperman DJ, Lavandera I, Gotor-Fernandez V, Gotor V, Taglieber A, Arends IWCE, Hollmann F (2013) Org Lett 15:180
114. Mifsud Grau M, van der Toorn JC, Otten LG, Macheroux P, Taglieber A, Zilly FE, Arends IWCE, Hollmann F (2009) Adv Synth Catal 351:3279

Adv Biochem Eng Biotechnol (2017) 158: 159–176
DOI: 10.1007/10_2016_48

Published online: 10 January 2017

Photosynthetic Microbial Fuel Cells

Joseph A. Laureanti and Anne K. Jones

Abstract This chapter presents the current state of research on bioelectrochemical systems that include phototrophic organisms. First, we describe what is known of how phototrophs transfer electrons from internal metabolism to external substrates. This includes efforts to understand both the source of electrons and transfer pathways within cells. Second, we consider technological progress toward producing bio-photovoltaic devices with phototrophs. Efforts to improve these devices by changing the species included, the electrode surfaces, and chemical mediators are described. Finally, we consider future directions for this research field.

Keywords Algae, Biophotovoltaic devices, Cyanobacteria, Extracellular electron transfer, Photosynthetic mechanisms, Photosynthetic microbial fuel cell

Contents

J.A. Laureanti and A.K. Jones (✉)
School of Molecular Sciences, Arizona State University, Tempe, AZ 85287, USA
e-mail: jonesak@asu.edu

Abbreviations

BES	Bio-electrochemical system
BPV	Bio-photovoltaic cell
CCCP	Carbonyl cyanide *m*-chlorophenyl hydrazone
DBMIB	2,5-Dibromo-3-methyl-6-isopropylbenzoquinone
DCMU	3-(3,4-Dichlorophenyl)-1,1-dimethylurea
EET	Extracellular electron transfer
ITO	Indium tin oxide
KCN	Potassium ferricyanide
MET	Microbial electrochemical technologies
MFC	Microbial fuel cell
PCP	Pentachlorophenol
PMA	Phenyl mercuric acetate
PMFC	Photosynthetic microbial fuel cell
PSII	Photosystem II

1 Introduction

The amount of solar energy arriving at the Earth in an hour is more than the annual human energy consumption. However, because sunlight is dilute and intermittent, the search for ways to capture, utilize, and store it efficiently has become the great scientific challenge of the twenty-first century. Blankenship and coauthors have demonstrated that short-term photosynthetic conversion yields of phototrophic microorganisms come within a factor of 2–3 of the best photovoltaic systems and have provided a compelling case for utilization of photosynthetic organisms in energy production and storage applications [1]. One approach is to use phototrophic organisms in microbial electrochemical technologies (MET) or bioelectrochemical systems (BES). This chapter describes recent progress in constructing and optimizing BES that employ an intact phototrophic organism at the cathode or anode and utilize the mechanisms of electron transfer from phototrophs to electrodes.

Microorganisms can be employed at either the anode or the cathode (or both) in a BES (Fig. 1). The most common type of MET is a system known as a microbial fuel cell (MFC) in which microorganisms are used as anode catalysts to oxidize an externally provided fuel and to transfer the resulting electrons to the voltaic system (Fig. 1a). This is usually coupled to the reduction of oxygen to water at the cathode. In the case where the fuel is a component of wastewater, the MFC can be used simultaneously both to treat the wastewater and to produce electricity [2]. Employing a microorganism instead at the cathode, that is, feeding electricity into the microorganism, results in a process known as electrosynthesis in which electrical energy is used to drive the production of a desired chemical (Fig. 1b) [3, 4]. Although not strictly speaking a microbial fuel cell, this is also an important application of bioelectrochemistry.

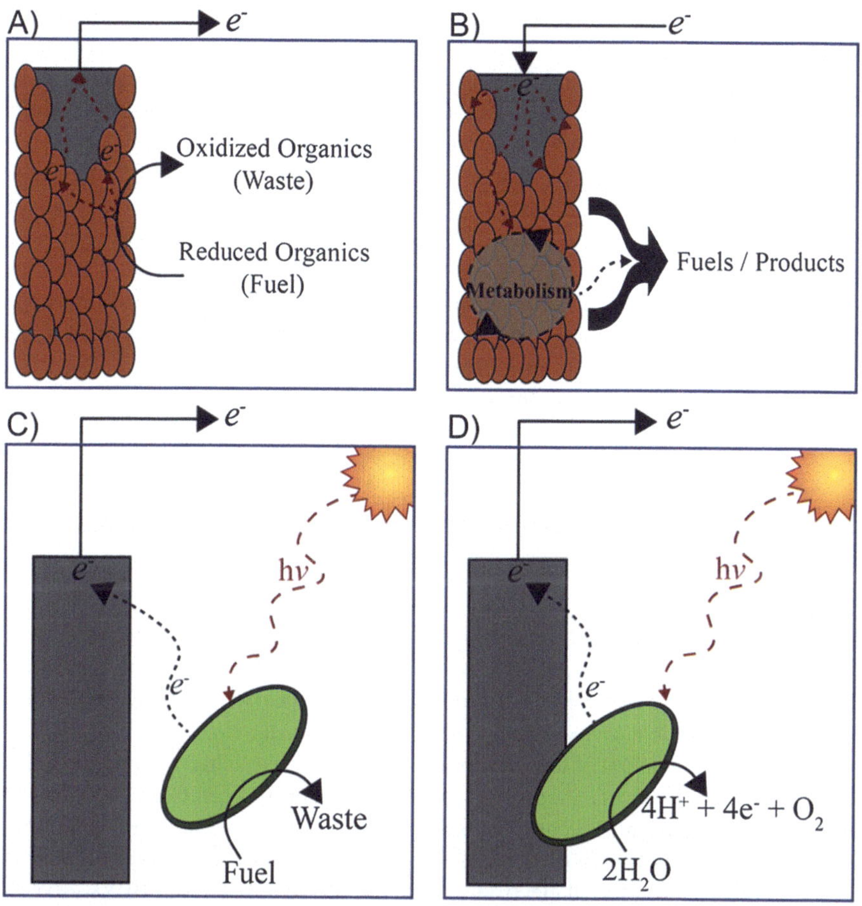

Fig. 1 Diagrams of bioelectrochemical systems. (**a**) Microbial fuel cell in which an anode-respiring bacterium produces current using electrons from a sacrificial, organic carbon source. (**b**) Electrosynthetic cell in which a microbe at the cathode uses electricity to drive chemical synthesis. (**c**) Photosynthetic microbial fuel cell in which a chemical electron donor is provided to an anoxygenic phototroph. (**d**) Biophotovoltaic system in which an oxygenic phototroph derived electrons at the anode from water

Photosynthetic microbial fuel cells (PMFC) and bio-photovoltaic cells (BPV) are two types of BES that employ at least one photosynthetic organism. In both cases, a phototroph is used at the anode to produce electrons. They differ in the source of those electrons. PMFCs utilize a sacrificial chemical fuel as the ultimate electron source (Fig. 1c). They may employ anaerobic phototrophs that are incapable of water-splitting [5]. On the other hand, a BPV uses an oxygenic photosynthetic organism at the bioanode to catalyze sunlight-driven photolysis of water in a traditional Z-scheme (Fig. 1d) and provides the resulting electrons to the voltaic system.

There is a natural comparison between light-harvesting BES and traditional photovoltaic devices. Because photosynthetic microorganisms are self-sustaining and inexpensive to culture, light-harvesting BES can offer tremendous advantages. Furthermore, generation of power from stored metabolites means microbial cells can also generate current under dark conditions, abrogating some of the challenge created by the intermittency of solar energy. However, development of BES is in its infancy. In particular, the efficiency of BPVs is relatively low, and understanding of the mechanisms that allow electrical communication between photosynthetic organisms and electrodes is less advanced than understanding of the analogous processes for other types of microorganisms. In this chapter, we first describe what is known of the mechanisms of electron transfer from phototrophs to extracellular substrates. We then consider the current state of the art in PMFC and BPV. We close by identifying promising future research directions.

2 Mechanisms of Electron Transfer Between Phototrophs and Electrodes

In microbial fuel cells, anode-respiring bacteria transfer electrons extracted from a fuel to the external electrode by moving them through respiratory pathways (Fig. 1a). A diverse array of phototropic species, especially cyanobacteria, also possess electrogenic activity, that is, the ability to transfer electrons from internal metabolism to an extracellular electrode. In many cases this activity is light-dependent [6]. However, it is often unclear whether the electrons are derived directly from photosynthesis or from some other metabolic process. In this section, we consider what is known of the electron transfer pathways that shuttle electrons from phototrophs to external electrodes. In so doing, we also see that BPVs serve as a unique tool to probe the interconnectivity of electron transfer pathways in phototrophs. Mechanistic understanding of extracellular electron transfer (EET) by phototrophs is relatively limited. Herein we describe what is known against a backdrop of the knowledge gleaned from the better-studied, anode-respiring *Shewanella* and *Geobacter* sp.

2.1 Indirect EET in Anode-Respiring Bacteria and Phototrophs

Studies of anode-respiring bacteria have identified two distinct mechanisms for EET: direct and indirect (Fig. 2). Indirect mechanisms are those that rely on a soluble redox mediator to shuttle electrons between the cell interior and the electrode. This shuttle can be either microbially produced or exogenously added. For example, *Shewanella oneidensis* MR-1 has been shown to reduce extracellular

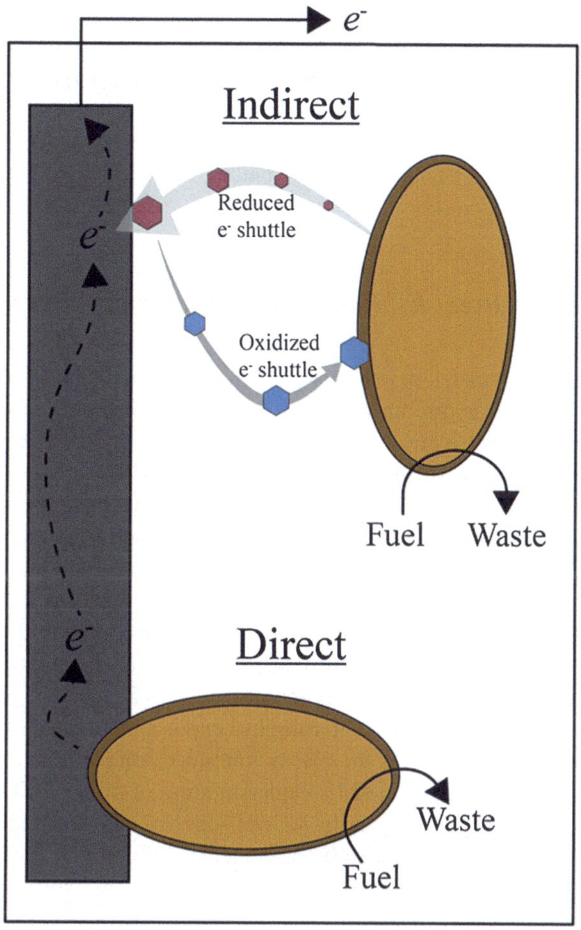

Fig. 2 Schematic depiction of direct and indirect mechanisms of extracellular electron transfer from a microbe to a solid electrode surface

minerals using endogenously produced and secreted flavins [7–11]. As much as 80% of EET by *Shewanella* is thought to be indirect [10]. Many phototrophs are also able to reduce exogenous mediators for indirect EET. The earliest BPVs relied almost exclusively on indirect transfer of electrons using such artificial mediators as viologens, napthoquinone derivatives [12, 13], and ferricyanide [14]. However, these mediators have largely fallen out of fashion because they offer significant problems for device scale-up and toxicity. As the natural electron acceptor of Photosystem II (PSII) is plastoquinone, Lemaître and colleagues hypothesized that quinones might serve as more effective mediators. They developed a fluorescence-based method to screen interactions with PSII rapidly and found that 2,6-dichlorobenzoquinone, 2,5-dichlorobenzoquinone, and *p*-phenylbenzoquinone can mediate electron transfer from the model green algae *Chlamydomonas reinhardtii* to electrodes [15]. Although there is no evidence that electrogenic phototrophs secrete flavins, they may naturally employ other endogenously-produced mediators to transfer reducing equivalents to electrodes. For example, hydrogen is a common photosynthetic product [16], and it can serve as a natural electron mediator between microbes and an anode [17–20]. Similarly, quinones have been hypothesized to serve as natural mediators of indirect EET in BPV [21, 22]. Several mediatorless BPVs have been described that may unwittingly rely on these endogenous mediators.

2.2 Direct EET in Anode-Respiring Bacteria

Direct transfer of electrons from microorganisms to an electrode is transfer that does not require an intervening chemical mediator. Instead, it occurs via physical contact between outer membrane proteins and/or conductive appendages and the extracellular substrate (Fig. 2). Direct EET by anode-respiring bacteria is hypothesized to employ conductive appendages that have been variously referred to in the literature as microbial nanowires or conductive pili [23, 24]. Recent studies have shown that the nanowires of *Shewanella* are not pili, as initially suggested, but rather outer membrane and periplasmic extensions [25]. Furthermore, *Shewanella* sp. are known to produce a collection of multiheme cytochromes that form multiple, interconnected electron transfer pathways from intracellular respiration and the quinone pool to the outer membrane [26–28]. These pathways are now relatively well-defined. In fact, heterologous expression of one, the MtrCAB pathway, confers on *Escherichia coli* the ability to reduce solid Fe_2O_3, an activity not present in the wild-type organism [29]. Understanding of direct EET by phototrophs, as we will see below, is not nearly so well-developed. Nonetheless, in the future, genetic manipulation may also be possible.

2.3 Direct EET by Phototrophs: Where Do Electrons Originate?

Initial reports of bio-nanowires in *S. oneidensis* MR-1 included preliminary evidence that the cyanobacterium *Synechocystis* sp. PCC 6803 can also produce electrically conductive nanowires under low carbon conditions [30]. However, this initial report has never been confirmed and additional evidence of conductive appendages has not been reported for this or any other phototroph. Thus it remains unclear whether phototrophs can participate in direct electron exchange with electrodes. Nonetheless, a broad range of phototrophs, including purple bacteria [20], cyanobacteria [6, 31–33], and eukaryotic algae [17, 34], have been shown to produce photocurrent at electrodes often without addition of an exogenous mediator [32, 35]. The question of the source of this current is further complicated by the anatomy of photosynthetic organisms and the location of the photosynthetic machinery. In algae, for example, photosynthesis occurs in the sub-cellular organelle known as a chloroplast, and the components are isolated from the electrode by a thick cell wall, making it likely that direct interaction between these components and an electrode is not a major component of photocurrent production. Similarly, the majority of cyanobacteria have two membrane systems: the cytoplasmic membrane and a collection of internal thylakoid membranes that house the photosynthetic and respiratory electron transport complexes [36]. Thus, transfer of photosynthetically-derived electrons to the cytoplasmic membrane must employ carriers that link the two membrane systems.

Observation of light-dependent current production by phototrophs in BPVs has led to the hypothesis that these electrons are derived not from biochemical oxidation of organic compounds but rather from light-driven water splitting. Herein, we first consider evidence that photocurrents in BPVs depend on oxygenic photosynthesis and water splitting. Then we explore hypotheses for the mechanisms of transport of photosynthetically-derived reducing equivalents to the cell surface.

Figure 3 shows the prototypical Z-scheme of oxygenic photosynthesis, highlighting site-specific inhibitors that can be used to probe the pathways of electrons from the photosystems to the cell exterior. PSII is the site of water oxidation, and Photosystem I (PSI) used to generate reductants for other cellular processes. The two photosystems are linked via the diffusing carrier plastoquinone

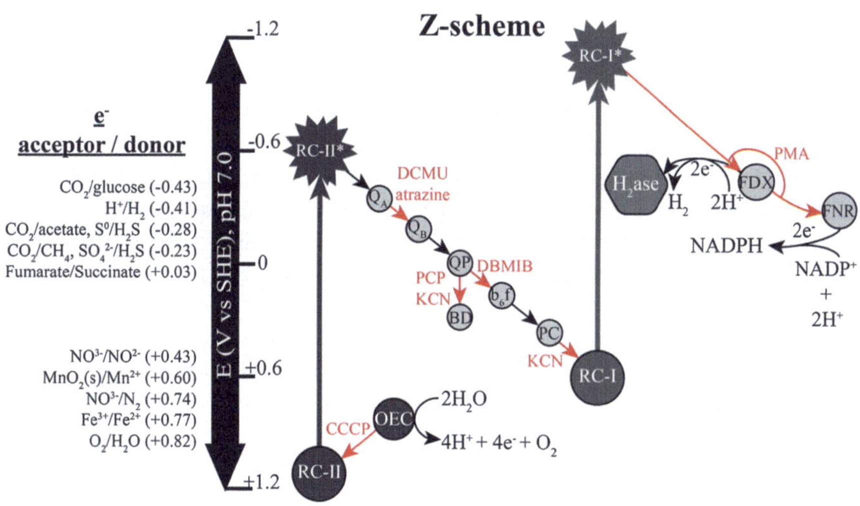

Fig. 3 Photosynthetic Z-scheme with reduction potentials of intermediates of the photosynthetic electron transport chain. Reduction potentials of common metabolic electron sources/sinks (*left*) are quoted relative to the standard hydrogen electrode, SHE, at pH 7.0 and are shown next to an energy level diagram oriented with the most reducing species at the *top*. *Red arrows* designate the inhibition of electron transfer using an exogenous chemical. Inhibitor abbreviations can be found in the abbreviation table. Z-scheme abbreviations are as follows: *RC-II* type-II reaction center, *OEC* oxygen evolving complex, *RC-I* type-I reaction center, Q_A quinone A, Q_B quinone b, *QP* quinone pool, *BD* cytochrome *bd* quinol oxidase, b_6f cytochrome b_6f complex, *PC* plastocyanin, *FDX* ferredoxin, *FNR* ferredoxin-NADP$^+$ reductase, H_2ase proton reducing hydrogenase

and the cytochrome b_6f complex. Photosynthetic electron flow is also linked to other metabolic processes such as respiration via soluble carriers. This can make it particularly difficult to ascertain the source of electrons.

Baskakov and colleagues suggested that the electrogenic activity of cyanobacteria is a means to protect the plastoquinone pool from overreduction at high light intensity [37]. Similarly, Freguia and coworkers have shown that cyanobacteria-dominated biofilms only produced photocurrent at solid electrodes when stressed with limited CO_2 or high light levels [38]. These "light stress" hypotheses have fueled the idea that EET by phototrophs is largely the result of excess electrons from photosynthesis, and a number of mechanistic studies explore this idea. Photosynthetic chemical inhibitors have been used to disrupt electron flow in BPVs employing cyanobacteria as a means to trace the electron transfer pathway. Pisciotta and coworkers used the PSII inhibitor 3-(3,4-dichloro-phenyl)-1,1,-dimethylurea (DCMU) to demonstrate that in *Lyngbya* and *Nostoc* electrons transferred to extracellular electron acceptors originate from the water splitting function of PSII [37]. The authors also concluded that plastoquinone is a key carrier in the pathways because addition of 2,5-dibromo-3-methyl-6-isopropyl benzoquinone (DBMIB), an inhibitor that prevents interactions between plastoquinones and cytochrome b_6f, enhanced electrogenic activity. Because oxidation of plastoquinol by cytochrome b_6f has been hypothesized to be a rate-limiting step in photosynthetic electron transfer [39], these results suggest that EET may serve phototrophs as a shunt to dissipate extra photosynthetic reducing equivalents. On the other hand, Bombelli and coworkers used analogous inhibitor studies to suggest that photosynthetically-derived electrons detected in a BPV employing *Synechocystis* are derived from the reducing end of PSI [14]. In a complementary approach, Cereda and coworkers demonstrated that circa 80% of extracellular electrons produced by *Synechocystis* in a BPV are derived from the water splitting activity of PSII by measuring current produced by a mutant strain lacking this activity [32]. This result suggests that, although the majority of photocurrent arises from photosynthetic water splitting, a considerable minority of electrons are unaccounted for photosynthetically and likely derived from respiratory electron transport. Thus it is likely that most photocurrent produced in BPV is derived from water splitting, but there is cross-talk with other metabolic pathways. This situation is not unlike that of *Shewanella* sp. in which many metabolic pathways compete with the electrode for electrons.

2.4 Transfer of Electrons from the Site of Photosynthesis to the Cell Surface

As photosynthetic membranes are typically localized in the interior of phototrophic cells, EET requires a pathway for reducing equivalents to be transferred from the site of production to the cell surface. Ferredoxins and NADPH both connect a multitude of pathways in algae and cyanobacteria and have been considered possible candidates for electron shuttles. In this section, we consider studies that address this hypothesis and efforts to enhance EET by introducing non-native electron carriers.

There is evidence suggesting that the thylakoid and cytoplasmic membranes are connected in *Synechocystis* [40]. However, it is unclear whether their electron transport chains are also connected. Howe and coworkers have hypothesized that proteins or mediators transfer reducing equivalents from the thylakoid membranes to the cytoplasmic membrane. Candidate carriers include the eight putative ferredoxins encoded by *Synechocystis*, soluble cytochromes, plastoquinol, and NADPH generated by ferredoxin-$NADP^+$ reductase. From the cytoplasmic membrane, electrons may then be transferred to ferredoxin or cytochromes in the periplasm to effect EET [41].

A similar mechanism may operate in algae. Transgenic *Chlamydomonas reinhardtii* strains overexpressing *PETF* and *FDX5*, the genes encoding two of the native ferredoxins for this model algae, have lower levels of reactive oxygen species, are more tolerant to heat and salt stress, accumulate more starch, and generate up to five times more power density in a PMFC [34]. Although these initially sound like unrelated phenotypes, ferredoxin is thought to be a central linker between photosynthetic electron production and these downstream reductive processes. Some of these pathways may serve as protection from the reactivity of energetic electrons produced under high level light conditions, and the results suggest that activity in PMFCs may be enhanced by eliminating competing pathways and forcing electrons toward the extracellular acceptor. Alternatively, it is possible that increased starch or simply biomass is necessary for enhanced activity in the PMFC. This biomass may be converted to NADPH which then feeds into the EET pathway.

Power densities in BPVs are approximately two orders of magnitude lower than those in MFCs employing anode-respiring bacteria such as *Geobacter* or *Shewanella*. This suggests that the natural electrogenic activity of phototrophs may be substantially lower than that of the anode-respiring bacteria. To improve this natural activity, Ramasamy and coworkers have heterologously overexpressed the *Geobacter* outer membrane cytochrome S (OmcS), a key component in EET, in the cyanobacterium *Synechococcus elongatus* PCC 7942. The variant cyanobacterium produced nine times more current than wild-type [42]. This preliminary result suggests that genetic engineering may offer significant opportunities to enhance the efficiency of EET by phototrophs. At the moment, the main challenge is identifying the genetically-tractable phototrophs most likely to be successfully deployed in BES.

3 Engineering of PMFC and BPV

The first PMFC was reported in 1964 by Berk and Canfield [18]. The anode of this cell consisted of planktonic *Rhodospirillum rubrum* cells fed by malate. The cells grow photoheterotrophically with concomitant production of hydrogen which is oxidized at the platinum anode. The resulting electrons are used at the cathode to reduce oxygen to water. The oxygen was produced by a photosynthetic biofilm of the marine algae *Oscillatoriaceae*. The algae grow photoautotrophically via fixation of atmospheric CO_2. The device could operate for several days as long as nutrients were regularly replenished, and conversion efficiency of incident solar irradiation to electrical power was 0.1–0.2%. Although this seems low, overall photosynthetic yields for converting solar irradiation to chemical energy range from <1% for agricultural crops to <3% for relatively productive algae. So this initial PMFC effort was within an order of magnitude of theoretical yields. Importantly, current devices have similar yield.

Following the work of Berk and Canfield, there have been numerous attempts to improve the power outputs of PMFCs. McCormick and coworkers reviewed efforts to engineer BPVs, and Fig. 4 shows evolution of performance over the last 30 years. Unfortunately, there has been no consistent improvement. To date, the greatest current density reported, 1.5 mA/cm [2], is from a BPV employing *Chlamydomonas reinhardtii* at a bioanode. Similar to the original Berk and Canfield cell, this BPV also relies on hydrogen produced by the microbe, in this latter case a green algae, to be oxidized at the platinum anode [17]. In light of scalability concerns, there is a trend away from platinum electrodes. Carbon electrode materials have become the standard in MFCs, and current densities as high as only fourfold lower than those on platinum have been reported by Yagishita and coworkers for a carbon-based BPV. As hydrogen is not efficiently oxidized at carbon substrates, carbon-based devices require alternative electron mediators. Thus Yagishita employed not just a different electrode surface but also a different redox mediator. He chose the soluble redox mediator 2-hydroxy-1,4-naphthoquinone in concert with *Synechococcus* PCC UTEX 2380 at the anode [13]. Addition of an exogenous mediator, however, also introduces concerns about scale-up and toxicity. Ideally, devices would be produced using only cheap, renewable, non-toxic materials. However, such a system with good performance metrics has not yet been created. Selection of organisms and electrode/mediator are two active areas of research being pursued to improve devices. Each is considered in turn below.

3.1 Identifying Organisms for Use in BPV

BPV devices have been constructed using both single species and microbial consortia as well as planktonic and benthic, biofilm forming, species [43–45]. Metagenomic evaluation of biofilm consortia in anerobic experiments demonstrates that these communities contain largely cyanobacteria as their photosynthetically active members [46]. Nonetheless, studies in pure culture have shown that phototrophs including purple bacteria [20], cyanobacteria [6, 13, 31–33], and eukaryotic algae [34, 47] can produce photocurrent at an electrode, but there is no consensus as to which are the best phototrophs for use in BES.

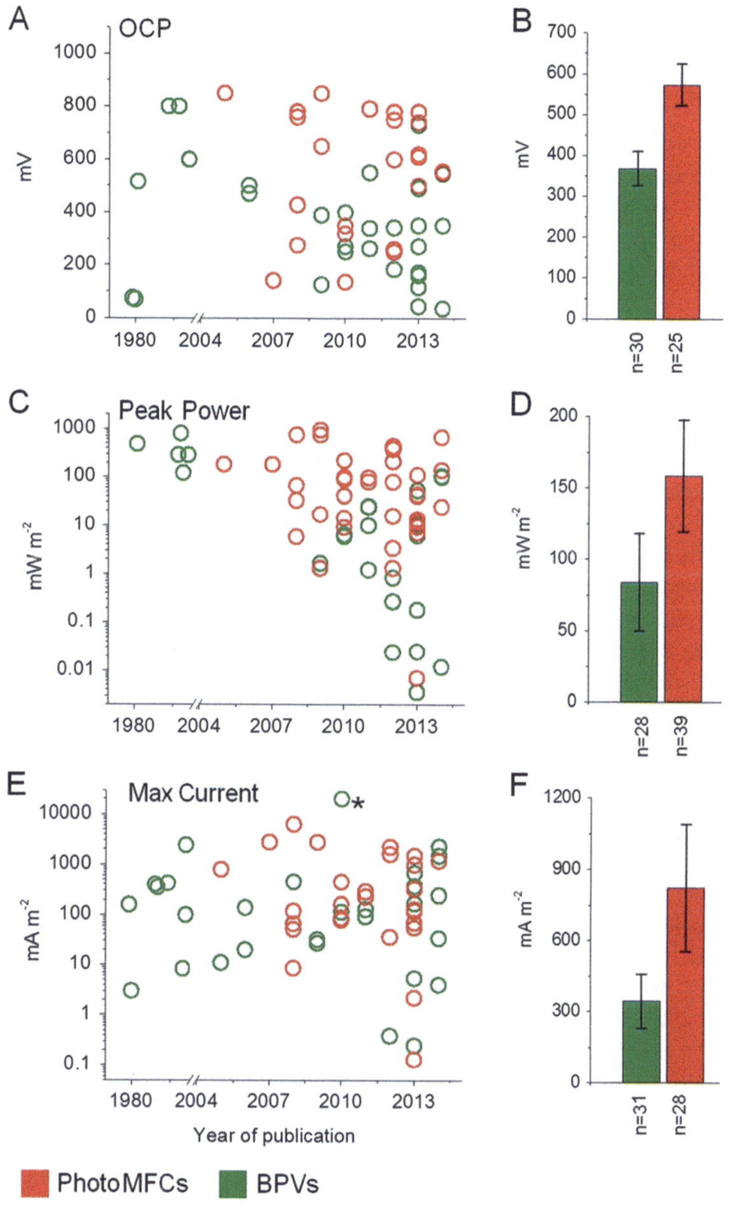

Fig. 4 Overview of BPV (*green*) and PMFC (*red*) performances. OCP stands for open circuit potential. (**b**, **d**, **f**) Averages for (**a**, **c**, **e**), respectively in which *n* indicates the number of studies considered. *Asterisk* indicates that point was not included in the averages. Reproduced with permission from A. J. McCormick, P. Bombelli, R. W. Bradly, R. Thorne, T. Wenzel, and C. J. Howe, *Energy Environ. Sci.*, 2015, **8**, 1092 – Published by the Royal Society of Chemistry

There are very few studies that report comparison of different species or consortia in the same electrochemical apparatus. Three key examples stand out. First, McCormick and coworkers compared four green algae and cyanobacteria, *Chlorella vulgaris*, *Dunaliella tertiolecta*, *Synechocystis* sp. PCC 6803, and *Synechococcus* sp. WH 5701, in a mediatorless BPV [35]. They concluded that all species demonstrated light-dependent activity. Second, Packer and co-workers designed a cost-effective cell to screen phototrophs (Fig. 5). They used it to evaluate the performance of 25 different benthic cyanobacteria and algal isolates from Antarctica and New Zealand. The same genera of cyanobacteria, *Pseudoanabaena*, *Leptolyngbya*, *Chroococales*, *M. vaginatus*, *Nostoc*, and *Phormidium* were consistently represented in highly functioning BPV. However, electrogenic performance does not strictly correlate with the genus. For example, highly electrogenic species from Antarctic samples were often from genera with modest or poor activity in samples from New Zealand [48]. Third, Fisher and coworkers used a similar approach to evaluate electrogenic activity from four algal strains. Interestingly, this group started from 16 strains, but 12 were eliminated before electrochemical testing for poor biofilm formation or poor overall photosynthetic performance. The power output by the four strains tested, *Synechococcus elongatus* (UMACC 105), *Chlorella vulgaris* UMACC 051, *Clorella* sp. (UMACC 313), and *Spirulina platensis* (UMACC 159), varied by over an order of magnitude [49]. This highlights the importance of screening more environmental samples for electrogenic activity to allow better matching of microbial properties to desired applications.

3.2 Materials for Electrodes

Similar to the impact of the microorganism employed, the role of the anode material in BPV performance has not been systematically evaluated [50]. Carbon anodes are de rigueur in microbial fuel cells, but preliminary evidence suggests that other materials may be highly advantageous for BPVs. For example, because BPVs require the microorganism to absorb light, there are obvious advantages to using transparent electrodes made of materials such as conducting metal oxides. McCormick and coworkers compared the performance of the fresh-water, filamentous cyanobacterium *Pseudanabaene limnetica* in a BPV on four different anode surfaces: indium tin oxide coated polyethylene tetraphthlate (ITO), stainless steel, glass coated with conductive polymer, and carbon paper. Carbon had the lowest power output in this system. The best performing materials were ITO and stainless steel [51]. Cameron and coworkers showed that *Chlorella vulgaris* biofilms grew more successfully on a porous fluorine-doped tin oxide coated titanium dioxide anode. The power density from this system was 16 times higher than a comparable carbon anode. Similar results were obtained with planar FTO coated glass, but the biofilms were less stable [52].

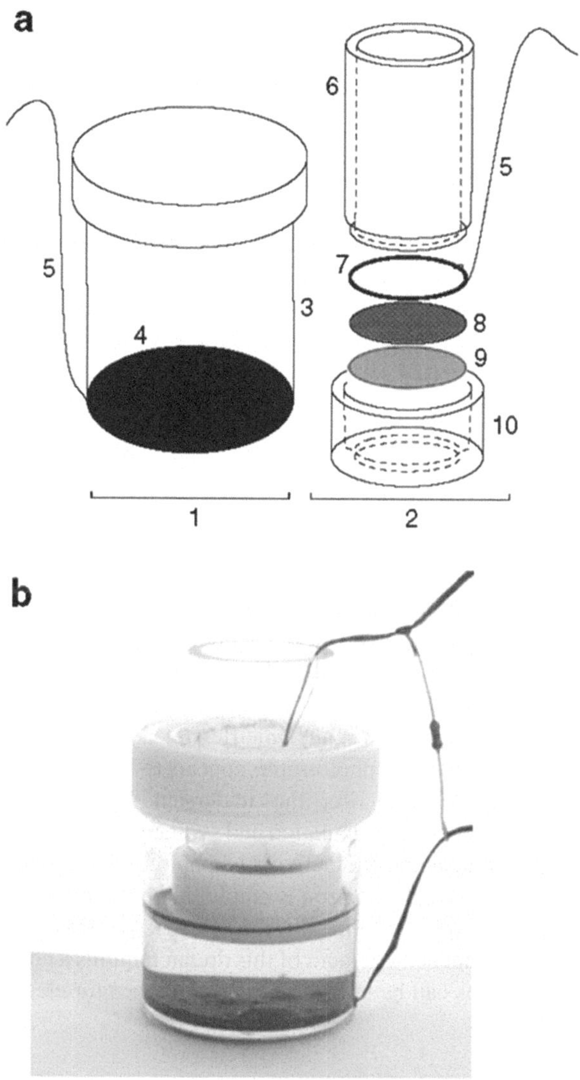

Fig. 5 (**a**) Schematic of PMFC and assembly detail. (**b**) Photograph of a PMFC containing a 1-month culture of *Paulschulzia pseudovolvox*. The anode chamber (*1*) is comprised of a polystyrene sample pottle (*3*) coated with Wire Glue™ (*4*). The cathode assembly insert (*2*) is comprised of a clear polycarbonate barrel (*6*) and a machines polyethylene cap (*10*). The cap fits tightly over the barrel, sealing a stack of Ultrex cathode exchange membrane (*9*), carbon cloth containing 10% platinum (*8*, Pt face down) and a titanium ring (*7*) to act as a charge collector which fits into lip machines on the end of *6*. Threads from a woven carbon fiber cloth were used as leads for the fuel cell (*5*); for the cathode, this was simply wrapped around the titanium ring before being pressed into place. The anode chamber was glued around the base of the chamber, acting as the charge collector. Journal of Applied Phycology, A cost-effective microbial fuel cell to detect and select for photosynthetic electrogenic activity in algae and cyanobacteria, 26, 2014, 15–23, V. M. Luimstra, S.-J. Kennedy, J. Güttler, S. A. Wood, D. E. Williams, and M. A. Packer. © Springer Science + Business Media Dordrecht 2013. With permission of Springer

Conductive polymer electrodes have also been evaluated as a means to transfer electrons more effectively from the microorganisms to the electrode. For example, Baskakov and coworkers employed nanostructured fibrillar polypyrrole as an anode material to harvest electrons from photosynthetic biofilms. Relative to measurements without the polymer, power density increased 450% [53]. Similarly, Gorton and coworkers have shown that Os-polymers can be used to mediate electron transfer from the green algae *Paulschulzia pseudovolvox* to graphite with current densities an order of magnitude higher than unmediated transfer. Interestingly, the current density can be further improved by an order of magnitude to 6.97 $\mu A\ cm^{-2}$ by inclusion of the diffusing redox mediator benzoquinone [54]. These results suggest that development of systems that combine multiple mediators with transparent electrodes may prove an exciting future research direction.

4 Conclusions

Creation of bioelectrochemical systems that employ phototrophs is a rapidly growing field which benefits from enhanced mechanistic understanding. Recent work shows that electrons can be derived from photosynthetic water splitting, but the pathway of the electron from PSII to the cell exterior remains unclear. Elucidating these details and mapping the interconnectivity between cellular electron transfer pathways may help improve performance metrics for devices. On the other hand, the native electrogenic activity of phototrophs appears to be much lower than that of model anode-respiring bacteria. As the understanding of the pathways in chemotrophs expands, genetic manipulation and synthetic biology is also becoming a viable approach to enhancing the activity of phototrophs. Finally, perhaps the most exciting application on the horizon is employing phototrophs as bio-factories for solar-driven production of chemicals from electricity. This chapter describes EET from phototrophs, but achievement of this dream requires a concerted effort to understand how electrons can be transferred into microbes for electrosynthesis.

References

1. Blankenship RE et al. (2011) Comparing photosynthetic and photovoltaic efficiencies and recognizing the potential for improvement. Science 332:805–809
2. Logan BE, Rabaey K (2012) Conversion of wastes into bioelectricity and chemicals by using microbial electrochemical technologies. Science 337:686–690
3. Nevin KP, Woodard TL, Franks AE, Summers ZM, Lovley DR (2010) Microbial electrosynthesis: feeding microbes electricity to convert carbon dioxide and water to multicarbon extracellular organic compounds. Mbio 1
4. Rabaey K, Rozendal RA (2010) Microbial electrosynthesis – revisiting the electrical route for microbial production. Nat Rev Microbiol 8:706–716

5. Chandra R, Modestra JA, Mohan SV (2015) Biophotovoltaic cell to harness bioelectricity from acidogenic wastewater associated with microbial community profiling. Fuel 160:502–512
6. Zou YJ, Pisciotta J, Billmyre RB, Baskakov IV (2009) Photosynthetic microbial fuel cells with positive light response. Biotechnol Bioeng 104:939–946
7. Brutinel ED, Gralnick JA (2012) Shuttling happens: soluble flavin mediators of extracellular electron transfer in *Shewanella*. Appl Microbiol Biotechnol 93:41–48
8. Okamoto A, Nakamura R, Nealson KH, Hashimoto K (2014) Bound flavin model suggests similar electron-transfer mechanisms in *Shewanella* and *Geobacter*. ChemElectroChem 1:1808–1812
9. Edwards MJ et al. (2015) Redox linked flavin sites in extracellular decaheme proteins involved in microbe-mineral electron transfer. Sci Rep 5
10. Kotloski NJ, Gralnick JA (2013) Flavin electron shuttles dominate extracellular electron transfer by *Shewanella oneidensis*. Mbio 4
11. Marsili E et al. (2008) *Shewanella* secretes flavins that mediate extracellular electron transfer. Proc Natl Acad Sci U S A 105:3968–3973
12. Ochiai H, Shibata H, Sawa Y, Shoga M, Ohta S (1983) Properties of semiconductor electrodes coated with living films of cyanobacteria. Appl Biochem Biotechnol 8:289–303
13. Yagishita T, Sawayama S, Tsukahara KI, Ogi T (1997) Effects of intensity of incident light and concentrations of *Synechococcus* sp. and 2-hydroxy-1,4-naphthoquinone on the current output of photosynthetic electrochemical cell. Solar Energy 61:347–353
14. Bombelli P et al. (2011) Quantitative analysis of the factors limiting solar power transduction by *Synechocystis* sp. PCC 6803 in biological photovoltaic devices. Energ Environ Sci 4:4690–4698
15. Longatte G et al. (2015) Evaluation of photosynthetic electrons derivation by exogenous redox mediators. Biophys Chem 205:1–8
16. Beer LL, Boyd ES, Peters JW, Posewitz MC (2009) Engineering algae for biohydrogen and biofuel production. Curr Opin Biotechnol 20:264–271
17. Rosenbaum M, Schroder U, Scholz F (2005) Utilizing the green alga *Chlamydomonas reinhardtii* for microbial electricity generation: a living solar cell. Appl Microbiol Biotechnol 68:753–756
18. Berk RS, Canfield JH (1964) Bioelectrochemical energy conversion. Appl Microbiol 12:102
19. Rosenbaum M, Schroder U, Scholz F (2005) In situ electrooxidation of photobiological hydrogen in a photobioelectrochemical fuel cell based on *Rhodobacter sphaeroides*. Environ Sci Technol 39:6328–6333
20. Cho YK et al. (2008) Development of a solar-powered microbial fuel cell. J Appl Microbiol 104:640–650
21. Newman DK, Kolter R (2000) A role for excreted quinones in extracellular electron transfer. Nature 405:94–97
22. Bradley RW, Bombelli P, Rowden SJL, Howe CJ (2012) Biological photovoltaics: intra- and extra-cellular electron transport by cyanobacteria. Biochem Soc Trans 40:1302–1307
23. El-Naggar MY et al. (2010) Electrical transport along bacterial nanowires from *Shewanella oneidensis* MR-1. Proc Natl Acad Sci U S A 107:18127–18131
24. Reguera G et al. (2005) Extracellular electron transfer via microbial nanowires. Nature 435:1098–1101
25. Pirbadian S et al. (2014) *Shewanella oneidensis* MR-1 nanowires are outer membrane and periplasmic extensions of the extracellular electron transport components. Proc Natl Acad Sci U S A 111:12883–12888
26. Alves MN et al. (2015) Characterization of the periplasmic redox network that sustains the versatile anaerobic metabolism of *Shewanella* oneidensis MR-1. Front Microbiol 6
27. Fonseca BM et al. (2013) Mind the gap: cytochrome interactions reveal electron pathways across the periplasm of *Shewanella oneidensis* MR-1. Biochem J 449:101–108
28. Sturm G et al. (2015) A dynamic periplasmic electron transfer network enables respiratory flexibility beyond a thermodynamic regulatory regime. ISME J 9:1802–1811

29. Jensen HM et al. (2010) Engineering of a synthetic electron conduit in living cells. Proc Natl Acad Sci U S A 107:19213–19218
30. Gorby YA et al. (2006) Electrically conductive bacterial nanowires produced by *Shewanella oneidensis* strain MR-1 and other microorganisms. Proc Natl Acad Sci U S A 103:11358–11363
31. Pisciotta JM, Zou Y, Baskakov IV (2010) Light-dependent electrogenic activity of cyanobacteria. PLoS One 5
32. Cereda A et al. (2014) A bioelectrochemical approach to characterize extracellular electron transfer by Synechocystis sp. PCC 6803. PLoS One 9
33. Sekar N, Umasankar Y, Ramasamy RP (2014) Photocurrent generation by immobilized cyanobacteria via direct electron transport in photo-bioelectrochemical cells. Phys Chem Chem Phys 16:7862–7871
34. Huang LF, Lin JY, Pan KY, Huang CK, Chu YK (2015) Overexpressing ferredoxins in *Chlamydomonas reinhardtii* increase starch and oil yields and enhance electric power production in a photo microbial fuel cell. Int J Mol Sci 16:19308–19325
35. McCormick AJ et al. (2011) Photosynthetic biofilms in pure culture harness solar energy in a mediatorless bio-photovoltaic cell (BPV) system. Energ Environ Sci 4:4699–4709
36. Mullineaux CW (2014) Co-existence of photosynthetic and respiratory activities in cyanobacterial thylakoid membranes. Biochim Biophys Acta 1837:503–511
37. Pisciotta JM, Zou YJ, Baskakov IV (2011) Role of the photosynthetic electron transfer chain in electrogenic activity of cyanobacteria. Appl Microbiol Biotechnol 91:377–385
38. Darus L, Ledezma P, Keller J, Freguia S (2016) Marine phototrophic consortia transfer electrons to electrodes in response to reductive stress. Photosynth Res 127:347–354
39. Trubitsin BV et al. (2005) EPR study of electron transport in the cyanobacterium *Synechocystis* sp. PCC 6803: oxygen-dependent interrelations between photosynthetic and respiratory electron transport chains. Biochim Biophys Acta 1708:238–249
40. Pisareva T et al. (2011) Model for membrane organization and protein sorting in the cyanobacterium *Synechocystis* sp. PCC 6803 inferred from proteomics and multivariate sequence analyses. J Proteome Res 10:3617–3631
41. Lea-Smith DJ, Bombelli P, Vasudevan R, Howe CJ (2016) Photosynthetic, respiratory and extracellular electron transport pathways in cyanobacteria. Biochim Biophys Acta 1857:247–255
42. Sekar N, Jain R, Yan Y, Ramasamy RP (2016) Enhanced photo-bioelectrochemical energy conversion by genetically engineered cyanobacteria. Biotechnol Bioeng 113:675–679
43. Cao XX, Huang X, Boon N, Liang P, Fan MZ (2008) Electricity generation by an enriched phototrophic consortium in a microbial fuel cell. Electrochem Commun 10:1392–1395
44. Darus L, Lu Y, Ledezma P, Keller J, Freguia S (2015) Fully reversible current driven by a dual marine photosynthetic microbial community. Bioresour Technol 195:248–253
45. Badalamenti JP, Torres CI, Krajmalnik-Brown R (2013) Light-responsive current generation by phototrophically enriched anode biofilms dominated by green sulfur bacteria. Biotechnol Bioeng 110:1020–1027
46. Nishio K, Hashimoto K, Watanabe K (2010) Light/electricity conversion by a self-organized photosynthetic biofilm in a single-chamber reactor. Appl Microbiol Biotechnol 86:957–964
47. Fu CC, Hung TC, Wu WT, Wen TC, Su CH (2010) Current and voltage responses in instant photosynthetic microbial cells with *Spirulina platensis*. Biochem Eng J 52:175–180
48. Luimstra VM et al. (2014) A cost-effective microbial fuel cell to detect and select for photosynthetic electrogenic activity in algae and cyanobacteria. J Appl Phycol 26:15–23
49. Ng FL, Phang SM, Periasamy V, Yunus K, Fisher AC (2014) Evaluation of algal biofilms on indium tin oxide (ITO) for use in biophotovoltaic platforms based on photosynthetic performance. PLoS One 9:13
50. Schneider K, Thorne RJ, Cameron PJ (2016) An investigation of anode and cathode materials in photomicrobial fuel cells. Philos Trans R Soc A Math Phys Eng Sci 374

51. Bombelli P et al. (2012) Surface morphology and surface energy of anode materials influence power outputs in a multi-channel mediatorless bio-photovoltaic (BPV) system. Phys Chem Chem Phys 14:12221–12229
52. Thorne R et al. (2011) Porous ceramic anode materials for photo-microbial fuel cells. J Mater Chem 21:18055–18060
53. Zou YJ, Pisciotta J, Baskakov IV (2010) Nanostructured polypyrrole-coated anode for sun-powered microbial fuel cells. Bioelectrochemistry 79:50–56
54. Hasan K et al. (2015) Photoelectrochemical wiring of Paulschulzia pseudovolvox (algae) to osmium polymer modified electrodes for harnessing solar energy. Adv Energy Mater 5

Index

The manufacturer's authorised representative in the EU is Springer Nature Customer Service Centre GmbH, Europaplatz 3, 69115 Heidelberg, Germany. If you have any concerns regarding our products, please contact ProductSafety@springernature.com

Printed and bound by CPI Group (UK) Ltd, Croydon, CR0 4YY
25/09/2026
02222859-0019